EVEN MORE Picture-Perfect SCIENCE Lessons, K–5

Using Children's Books to Guide Inquiry

Even More Picture-Perfect SCIENCE Lessons, K–5

Using Children's Books to Guide Inquiry

by Emily Morgan and Karen Ansberry

Arlington, Virginia

Claire Reinburg, Director
Jennifer Horak, Managing Editor
Andrew Cooke, Senior Editor
Amanda O'Brien, Associate Editor
Wendy Rubin, Associate Editor
Amy America, Book Acquisitions Coordinator

Art and Design
Will Thomas Jr., Director
Linda Olliver, Cover, Interior Design, Illustrations

Printing and Production
Catherine Lorrain, Director

National Science Teachers Association
David L. Evans, Executive Director
David Beacom, Publisher

1840 Wilson Blvd., Arlington, VA 22201
www.nsta.org/store
For customer service inquiries, please call 800-277-5300.

18 17 16 15   6 5 4 3

Library of Congress Cataloging-in-Publication Data
Ansberry, Karen Rohrich, 1966-
Even more picture-perfect science lessons : using children's books to guide inquiry, K-5 / by Karen Ansberry and Emily Morgan.
pages cm
Includes index.
ISBN 978-1-935155-17-1
1. Science--Study and teaching (Elementary) 2. Picture books for children--Educational aspects. I. Morgan, Emily R. (Emily Rachel), 1973- II. Title.
LB1585.A568 2013
372.35044--dc23
2013001262

Cataloging-in-Publication Data are also available from the Library of Congress for the e-book.
e-ISBN: 978-1-938946-89-9

Contents

Preface

A class of fifth-grade students listen as their teacher reads *The Boy Who Harnessed the Wind*. This is the true story of William Kamkwamba, a 14-year-old African boy whose perseverance and ingenuity brought electricity and running water to his drought-ravaged village. With nothing but scraps from a junkyard and the knowledge he acquired from library books, William built a windmill and waited for the wind. Students sit in awe as the teacher reads the dramatic account of what happened next. "Like always, it came, first a breeze, then a gusting gale. The tower swayed and the blades spun round. With sore hands once slowed by hunger and darkness William connected wires to a small bulb, which flickered at first, then surged as bright as the sun. 'Tonga!' he shouted. 'I have made electric wind!'" The teacher asks the class, "How do you think William's windmill was able to light a lightbulb?" In a lesson inspired by this extraordinary story (Chapter 9, "Harnessing the Wind"), students discover the answer to this question by first investigating how a simple handheld generator, the Dynamo Torch, transforms energy of motion into electrical energy. Students then build on this experience by reading an article about energy transformations and listening to a nonfiction read-aloud that explains how wind turbines produce electricity. Eventually, students develop explanations that explain how William's windmill was able to light a bulb. They elaborate on what they have learned by researching various energy resources. Through hands-on inquiries and high-quality readings and picture books, students learn difficult concepts about energy—all within the context of William's remarkable true story.

What Is Picture-Perfect Science?

This scenario describes how a children's picture book can help guide students through an engaging, hands-on inquiry lesson. *Even More Picture-Perfect Science Lessons, K–5* contains 20 science lessons for students in grades K through 5, with embedded reading comprehension strategies to help them learn to read and read to learn while engaged in inquiry-based science. To help you teach according to *A Framework for K–12 Science Education* (NRC 2012), the lessons are written in an easy-to-follow format for teaching inquiry-based science according to the BSCS 5E Instructional Model (Bybee 1997, used with permission from BSCS; see Chapter 4 for more information). This learning cycle model allows students to construct their own understanding of scientific concepts as they cycle through the following phases: engage, explore, explain, elaborate, and evaluate. Although *Even More Picture-Perfect Science Lessons* is primarily a book for teaching science, reading comprehension strategies and the Common Core State Standards for English Language Arts (Common Core ELA; NGA for Best Practices and CCSSO 2010) are embedded in each lesson. These essential strategies can be modeled while keeping the focus of the lessons on science.

Use This Book Within Your Science Curriculum

We wrote *Even More Picture-Perfect Science Lessons* to supplement, not replace, an existing science program. Although each lesson stands alone as a carefully planned learning cycle based on clearly defined science objectives, the lessons are intended to be integrated into a more complete unit of instruction in which concepts can be more fully developed. The lessons are not designed to be taught sequentially. We want you to use the lessons where appropriate within your school's current science curriculum to support, enrich, and extend it. And we want you to adapt the lessons to fit your school's curriculum, your students' needs, and your own teaching style.

Special Features

Ready-to-Use Lessons With Assessments

Each lesson contains engagement activities, hands-on explorations, student pages, suggestions for student and teacher explanations, opportunities for elaboration, assessment suggestions, and annotated bibliographies of more books to read on the topic. Assessments include poster sessions with rubrics, teacher checkpoint labs, and formal multiple-choice and extended-response questions.

Reading Comprehension Strategies

Reading comprehension strategies based on the book *Strategies That Work* (Harvey and Goudvis 2000) and specific activities to enhance comprehension are embedded throughout the lessons and clearly marked with an icon. Chapter 2 describes how to model these strategies while reading aloud to students.

Standards-Based Objectives

All lesson objectives are grade-level endpoints from *A Framework for K–12 Science Education* (NRC 2012) and are clearly identified at the beginning of each lesson. Because we wrote *Even More Picture-Perfect Science Lessons* for students in grades K–5, we used two grade ranges of the *Framework:* K–2 and 3–5. Chapter 5, "Connecting to the Standards," outlines the component ideas from the *Framework* and the grade band addressed for each lesson.

The lessons also incorporate the Common Core State Standards for English Language Arts. In a box titled "Connecting to the Common Core" you will find the Common Core ELA strand the activity addresses (e.g., reading, writing, speaking and listening, or language) as well as the grade level and standard number (e.g., K.9 or 5.1). You will see that writing assignments are specifically labeled with an icon.

Science as Inquiry

As we said, the lessons in *Even More Picture-Perfect Science Lessons* are structured as guided inquiries following the 5E Model. Guiding questions are embedded throughout each lesson and marked with an icon **?**. The questioning process is the cornerstone of good teaching. A teacher who asks thoughtful questions arouses students' curiosity, promotes critical-thinking skills, creates links between ideas, provides challenges, gets immediate feedback on student learning, and helps guide students through the inquiry process. Chapters 3 and 4 explore science as inquiry and the BSCS 5E Instructional Model, and each lesson includes an "Inquiry Place" box that suggests ideas for developing open inquiries.

References

Bybee, R. W. 1997. *Achieving scientific literacy: From purposes to practices.* Portsmouth, NH: Heinemann.

Harvey, S., and A. Goudvis. 2000. *Strategies that work: Teaching comprehension to enhance understanding.* York, ME: Stenhouse Publishers.

National Governors Association Center (NGA) for Best Practices, and Council of Chief State School Officers (CCSSO). 2010. *Common core state standards for English language arts and literacy.* Washington, DC: National Governors Association for Best Practices, Council of Chief State School.

National Research Council (NRC). 2012. *A framework for K–12 science education: Practices, crosscutting concepts, and core ideas.* Washington, DC: National Academies Press.

Children's Book Cited

Kamkwamba, W., and B. Mealer. 2012. *The boy who harnessed the wind.* New York: Dial Books for Young Readers.

Editors' Note

Even More Picture-Perfect Science Lessons builds on the texts of 31 children's picture books to teach science. Some of these books feature animals that have been anthropomorphized, such as a caterpillar that does magic tricks. While we recognize that many scientists and educators believe that personification, teleology, animism, and anthropomorphism promote misconceptions among young children, others believe that removing these elements would leave children's literature severely underpopulated. Furthermore, backers of these techniques not only see little harm in their use but also argue that they facilitate learning.

Because *Even More Picture-Perfect Science Lessons* specifically and carefully supports scientific inquiry—the "Amazing Caterpillars" lesson, for instance, teaches students how to weed out misconceptions by asking them to point out inaccurate information about caterpillars and butterflies in a storybook—we, like our authors, feel the question remains open.

Acknowledgments

We would like to dedicate this book to the memory of Sue Livingston, who opened our eyes to the power of modeling reading strategies in the content areas and for teaching us that every teacher is a reading teacher.

We appreciate the care and attention to detail given to this project by Jennifer Horak, Agnes Bannigan, Pat Freedman, and Claire Reinburg at NSTA Press.

And these thank-yous as well:

- To Linda Olliver for her "Picture-Perfect" illustrations
- To the staff and students of Mason City Schools, Cincinnati Public Schools, and Lebanon United Methodist Preschool and Kindergarten for field-testing lessons and providing "photo ops"
- To Jackie Anderson, Fliss LaRosa, Jeff Morgan, and Rhonda Vanderbeek for contributing photographs
- To Shannon Homoelle for sharing her expertise with the Common Core State Standards for English Language Arts
- To Don Kaufman and Cecilia Berg for giving us the opportunity to share Picture-Perfect Science as part of the GREEN Teachers Institute at Miami University in Oxford, Ohio
- To Bill Robertson for sharing his content knowledge

The contributions of the following reviewers are also gratefully acknowledged:

- Carol Collins
- Miriam Jean Dreher
- Christine Pappas

Contributors

Ideas and activities for the lessons in this book were contributed by the following talented, dedicated teachers. We thank them for their creativity, willingness to share, and the important work they do each day in their classrooms.

Jackie Anderson is a multiple disabilities teacher at Roselawn Condon School in Cincinnati, Ohio. Jackie contributed to Chapter 11, "Do You Know Which Ones Will Grow?"

Allyson Day is a sixth-grade math and science teacher at Monroe Elementary School in Monroe, Ohio. Allyson contributed to Chapter 7, "Float Your Boat."

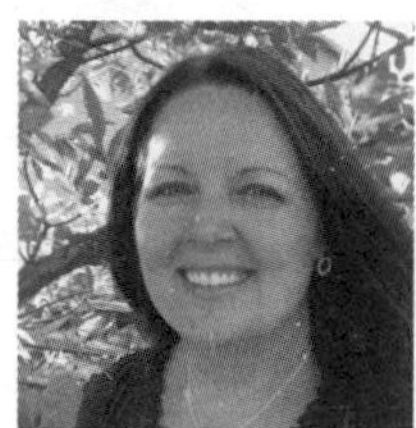

Missy Breuer is a fifth-grade teacher at Pattison Elementary School in Milford, Ohio. Missy contributed to Chapter 20, "Problem Solvers."

Jenny DeBord is a first-grade teacher at Monroe Primary School in Monroe, Ohio. Jenny contributed to Chapter 7, "Float Your Boat."

Tim Breuer is a seventh-grade teacher at Milford Junior High School in Milford, Ohio. Tim contributed to Chapter 18, "What Will the Weather Be?"

Jenny Doerflein is certified to teach art K–12. Jenny contributed to Chapter 11, "Do You Know Which Ones Will Grow?"

Katie Davis is a second-grade teacher at Mason Early Childhood Center in Mason, Ohio. Katie contributed to Chapter 14, "Ducks Don't Get Wet"; Chapter 16, "Fossils Tell of Long Ago"; and Chapter 17, "Reduce, Reuse, Recycle."

Karen Eads is a first-grade teacher at Sharpsburg Elementary School in Norwood, Ohio. Karen contributed to Chapter 6, "Freezing and Melting."

Maria Eshman is a first-grade teacher at Sharpsburg Elementary School in Norwood, Ohio. Maria contributed to Chapter 6, "Freezing and Melting."

Faye Harp is a teaching and learning consultant at Lakota Local Schools in West Chester, Ohio. Faye contributed to Chapter 12, "Seeds on the Move."

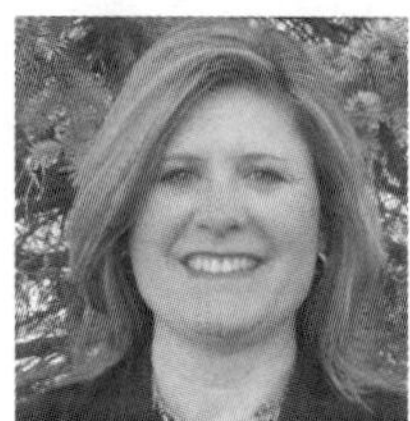

Kathy Gall is a second-grade teacher at Monroe Primary School in Monroe, Ohio. Kathy contributed to Chapter 7, "Float Your Boat."

Aubrey McCalla teaches first grade at Williamsburg Elementary School in Williamsburg, Ohio. Aubrey contributed to Chapter 11, "Do You Know Which Ones Will Grow?"

Michelle Gallite is a third-grade teacher at Western Row Elementary School in Mason, Ohio. Michelle contributed to Chapter 8, "The Wind Blew," and Chapter 15, "Amazing Caterpillars."

Colleen Phillips-Birdsong taught second grade for 11 years and now is a Reading Recovery teacher and reading specialist at Mercer Elementary School in Cincinnati. Colleen contributed to Chapter 10, "Sounds All Around"; Chapter 13, "Unbeatable Beaks"; and Chapter 19, "Sunsets and Shadows."

Lisa Haines is a title one reading teacher at Wilmington City Schools in Wilmington, Ohio. Lisa contributed to Chapter 11, "Do You Know Which Ones Will Grow?"

Katie Woodward is a second-grade teacher at Monroe Primary School in Monroe, Ohio. Katie contributed to Chapter 7, "Float Your Boat."

About the Authors

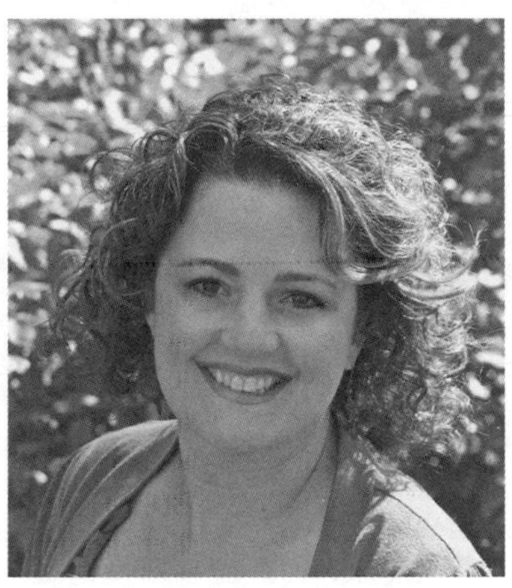

Emily Morgan is a former elementary science lab teacher for Mason City Schools in Mason, Ohio, and seventh-grade science teacher at Northridge Local Schools in Dayton, Ohio. She has a bachelor of science in elementary education from Wright State University and a master of science in education from the University of Dayton. She is also the author of the *Next Time You See* picture book series from NSTA Press. Emily lives in West Chester, Ohio, with her husband, son, and an assortment of animals.

Karen Ansberry is an elementary science curriculum leader and former fifth- and sixth-grade science teacher at Mason City Schools in Mason, Ohio. She has a bachelor of science in biology from Xavier University and a master of arts in teaching from Miami University. Karen lives in historic Lebanon, Ohio, with her husband, two sons, two daughters, and too many animals.

Emily and Karen, along with language arts consultant Sue Livingston, received a Toyota Tapestry grant for their Picture-Perfect Science grant proposal in 2002. Since then, they have enjoyed facilitating teacher workshops at elementary schools, universities, and professional conferences across the country. This is Emily and Karen's third book in the *Picture-Perfect Science Lessons* series.

Emily and Karen would like to dedicate this book to the memory of Sue Livingston.

About the Picture-Perfect Science Program

The Picture-Perfect Science program originated from Emily Morgan's and Karen Ansberry's shared interest in using children's literature to make science more engaging. In Emily's 2001 master's thesis study involving 350 of her third-grade science lab students at Western Row Elementary, she found that students who used science trade books instead of the textbook scored significantly higher on district science performance assessments than students who used the textbook only. Convinced of the benefits of using picture books to engage students in science inquiry and to increase science understanding, Karen and Emily began collaborating with Sue Livingston, Mason's elementary language arts curriculum leader, in an effort to integrate literacy strategies into inquiry-based science lessons. They received grants from the Ohio Department of Education (2001) and Toyota Tapestry (2002) to train all third- through sixth-grade science teachers, and in 2003 they also trained seventh- and eighth-grade science teachers with district support. The program has been presented at elementary schools, conferences, and universities nationwide.

For more information on Picture-Perfect Science teacher workshops, go to *www.pictureperfectscience.com.*

Lessons by Grade

Chapter	Grade	Picture Books
6	K–2	*Wemberly's Ice-Cream Star* *Why Did My Ice Pop Melt?*
7	3–5	*Toy Boat* *Captain Kidd's Crew Experiments With Sinking and Floating*
8	3–5	*The Wind Blew* *I Face the Wind*
9	3–5	*The Boy Who Harnessed the Wind* *Wind Energy: Blown Away!*
10	K–2	*What's That Sound?* *Sounds All Around*
11	K–2	*Do You Know Which Ones Will Grow?* *What's Alive?*
12	K–2	*Flip, Float, Fly: Seeds on the Move* *Who Will Plant a Tree?*
13	K–2	*Unbeatable Beaks* *Beaks!*
14	3–5	*Just Ducks!* *Ducks Don't Get Wet*
15	K–2	*Houdini the Amazing Caterpillar* *From Caterpillar to Butterfly* *The Very Hungry Caterpillar*
16	3–5	*Fossil* *Fossils Tell of Long Ago*
17	K–2	*The Three R's: Reuse, Reduce, Recycle* *Michael Recycle*
18	3–5	*Come On, Rain!* *What Will the Weather Be?*
19	3–5	*Twilight Comes Twice* *Next Time You See a Sunset*
20	3–5	*Now & Ben: The Modern Inventions of Benjamin Franklin* *Build It: Invent New Structures and Contraptions*

Why Read Picture Books in Science Class?

Think about a book you loved as a child. Maybe you remember the zany characters and rhyming text of Dr. Seuss classics like *One Fish Two Fish Red Fish Blue Fish* or the clever poems in Shel Silverstein's *Where the Sidewalk Ends*. Perhaps you enjoyed the page-turning suspense of Jon Stone's *The Monster at the End of This Book* or the fascinating facts found in Aliki's *Digging Up Dinosaurs*. You may have seen a little of yourself in *Where the Wild Things Are* by Maurice Sendak, *Ramona the Pest* by Beverly Cleary, or *Curious George* by H. A. Rey. Maybe your imagination was stirred by the colorful illustrations in Eric Carle's *The Very Hungry Caterpillar* or the stunning photographs in Seymour Simon's *The Moon*. You probably remember the warm, cozy feeling of having a treasured book like Arnold Lobel's *Frog and Toad Are Friends* or E. B. White's *Charlotte's Web* being read to you by a parent or grandparent. But chances are your favorite book as a child was *not* your fourth-grade science textbook. The format of picture books offers certain unique advantages over textbooks and chapter books for engaging students in a science lesson. More often than other books, fiction and nonfiction picture books stimulate students on

Teachers enjoy using picture books.

both the emotional and intellectual levels. They are appealing and memorable because children readily connect with the imaginative illustrations, vivid photographs, experiences and adventures of characters, engaging storylines, the fascinating information that supports them in their quest for knowledge, and the warm emotions that surround the reading experience.

What characterizes a picture book? We like what *Beginning Reading and Writing* says, "Picture books are unique to children's literature as they are defined by format rather than content. That is, they are books in which the illustrations are of equal importance as or more important than the text in the creation of meaning" (Strickland and Morrow 2000, p. 137). Because picture books are more likely to hold children's attention, they lend themselves to reading comprehension strategy instruction and to engaging students within an inquiry-based cycle of science instruction. "Picture books, both fiction and nonfiction, are more likely to hold our attention and engage us than reading dry, formulaic text. ... Engagement leads to remembering what is read, acquiring knowledge and enhancing understanding" (Harvey and Goudvis 2000, p. 46). We wrote the *Picture-Perfect Science Lessons* series so teachers can take advantage of the positive features of children's picture books by supplementing the traditional science textbook with a wide variety of high-quality fiction and nonfiction science-related picture books.

The Research

Context for Concepts

Literature gives students a context for the concepts they are exploring in the science classroom. Children's picture books, a branch of literature, have interesting storylines that can help students understand and remember concepts better than they would by using textbooks alone, which tend to present science as lists of facts to be memorized (Butzow and Butzow 2000). In addition, the colorful pictures and graphics in picture books are superior to many texts for explaining abstract ideas (Kralina 1993). As more and more content is packed into the school day and higher expectations are placed on student performance, it is critical for teachers to teach more in the same amount of time. Integrating curriculum can help accomplish this. The wide array of high-quality children's literature available can help you model reading comprehension strategies while teaching science content in a meaningful context.

More Depth of Coverage

Science textbooks can be overwhelming for many children, especially those who have reading problems. They often contain unfamiliar vocabulary and tend to cover a broad range of topics (Casteel and Isom 1994; Short and Armstrong 1993; Tyson and Woodward 1989). However, fiction and nonfiction picture books tend to focus on fewer topics and give more in-depth coverage of the concepts. It can be useful to pair an engaging fiction book with a nonfiction book to round out the science content being presented.

For example, the "What Will the Weather Be?" lesson in Chapter 18 features *Come On, Rain!* a beautifully descriptive story of a little girl waiting in the sizzling heat for the impending rain. It is paired with *What Will the Weather Be?* a nonfiction book that explains how various weather instruments are used to help meteorologists predict the weather. The expressive language and illustrations in *Come On, Rain!* hook the reader, and the book *What Will the Weather Be?* presents facts and background information. Together they offer a balanced, in-depth look at how changes in weather are predicted and how they affect us.

Improved Reading and Science Skills

Research by Morrow et al. (1997) on using children's literature and literacy instruction in the science program indicated gains in science as well as literacy. Romance and Vitale (1992) found significant improvement in both science and reading scores of fourth graders when the regular basal reading program was replaced with reading in science that correlated with the science

curriculum. They also found an improvement in students' attitudes toward the study of science.

Opportunities to Correct Science Misconceptions

Students often have strongly held misconceptions about science that can interfere with their learning. "Misconceptions, in the field of science education, are preconceived ideas that differ from those currently accepted by the scientific community" (Colburn 2003, p. 59). Children's picture books, reinforced with hands-on inquiries, can help students correct their misconceptions. Repetition of the correct concept by reading several books, doing a number of experiments, and inviting scientists to the classroom can facilitate a conceptual change in children (Miller, Steiner, and Larson 1996).

But teachers must be aware that scientific misconceptions can be inherent in the picture books. Although many errors are explicit, some of the misinformation is more implicit or may be inferred from text and illustrations (Rice 2002). This problem is more likely to occur within fictionalized material. Mayer's (1995) study demonstrated that when both inaccuracies and science facts are presented in the same book, children do not necessarily remember the correct information.

Scientific inaccuracies in picture books can be useful for teaching. Research shows that errors in picture books, whether identified by the teacher or the students, can be used to help children learn to question the accuracy of what they read by comparing their own observations to the science presented in the books (Martin 1997). Scientifically inaccurate children's books can be helpful when students analyze inaccurate text or pictures after they have gained understanding of the correct scientific concepts through inquiry experiences.

For example, in the "Amazing Caterpillars" lesson in Chapter 15, after observing live painted lady caterpillars go through metamorphosis and reading a nonfiction book about caterpillars and butterflies, students analyze Eric Carle's classic book *The Very Hungry Caterpillar* and then retell the book in a way that is scientifically accurate. This process requires students to think critically: They apply what they have learned to evaluate and correct the misinformation in the picture book. When correcting a fictional book, we always tell students that anything can happen in a fiction book and that it doesn't have to be scientifically accurate. But sometimes it is fun to look at a fictional story and point out what is true and what is not. In the Chapter 15 lesson, we use a quote from Eric Carle's website explaining why he chose to have the caterpillar in his story make a cocoon (which is actually the pupal stage of a moth) instead of a chrysalis.

Use With Upper Elementary Students

Even More Picture-Perfect Science Lessons is designed for students in grades K through 5. Although picture books are more commonly used with younger children, we have good reasons for recommending their use with upper elementary students. In *Strategies That Work* (2000), reading experts Harvey and Goudvis maintain that "The power of well-written picture books cannot be overestimated ... picture books lend themselves to comprehension strategy instruction at every grade level." The benefits of using picture books to teach science and reading strategies are not reserved for younger children. We have found them effective for engaging students, for guiding scientific inquiry, and for teaching comprehension strategies to students in kindergarten through eighth grade. We believe that the wide range of topics, ideas, and genres found in picture books reaches all readers, regardless of their ages, grades, reading levels, or prior experiences.

Selection of Books

Each lesson in *Even More Picture-Perfect Science Lessons* focuses on grade-level endpoints from *A Framework for K–12 Science Education* (NRC 2012). We selected fiction and nonfiction children's picture books that closely relate to these standards. An annotated "More Books to Read"

section is provided at the end of each lesson. If you would like to select more children's literature to use in your science classroom, try the Outstanding Science Trade Books for Students K–12 listing, a cooperative project between the National Science Teachers Association (NSTA) and the Children's Book Council (CBC). The books are selected by a book review panel appointed by NSTA and assembled in cooperation with CBC. Each year a new list is featured in the March issue of NSTA's elementary school teacher journal *Science and Children.* See *www.nsta.org/ostbc* for archived lists.

When you select children's picture books for science instruction, you should consult with a knowledgeable colleague who can help you check them for errors or misinformation. You might talk with a high school science teacher, a retired science teacher, or a university professor. To make sure that the books are developmentally appropriate or lend themselves to a particular reading strategy you want to model, you could consult with a language arts specialist.

Finding the *Picture-Perfect* Books

Each activity chapter includes a "Featured Picture Books" section with titles, author and illustrator names, publication details, and summaries of the books. The years and publisher names listed are for the most recent editions available—paperback whenever possible—as of the printing of *Even More Picture-Perfect Science Lessons.*

All of the trade books featured in the lessons in this book are currently in print and can be found at your local bookstore, online retailer, or library. All of the picture books—including previously hard-to-find and out-of-print books, such as *Unbeatable Beaks*—are available at *www.nsta.org/store*. There you can also buy *all* of the *Even More Picture-Perfect Science Lessons* books in one handy collection at a reduced cost; you can also buy ClassPacks, which contain the materials you need to do each lesson, at the NSTA online store.

Considering Genre

Considering genre when you determine how to use a particular picture book within a science lesson is important. Donovan and Smolkin (2002) identify four different genres frequently recommended for teachers to use in their science instruction: story, non-narrative information, narrative information, and dual purpose. *Even More Picture-Perfect Science Lessons* identifies the genre of each featured book at the beginning of each lesson. Summaries of the four genres, a representative picture book for each genre, and suggestions for using each genre within the BSCS 5E learning cycle we use follow. (The science learning cycle known as the BSCS 5E Instructional Model is described in detail in Chapter 4.)

Storybooks

Storybooks center on specific characters who work to resolve a conflict or problem. The major purpose of stories is to entertain, not to present factual information. The vocabulary is typically commonsense, everyday language. An engaging storybook can spark interest in a science topic and move students toward informational texts to answer questions inspired by the story. For example, the "Freezing and Melting" lesson in Chapter 6 uses *Wemberly's Ice-Cream Star*, a story about a little mouse who wants to share her ice cream with a friend, so she waits for it to melt into "ice cream soup." The charming story hooks the learners and engages them in explorations with solids and liquids.

Scientific concepts in stories are often implicit, so teachers must make the concepts explicit to students. As we mentioned, be aware that storybooks often contain scientific errors, either explicit or implied by text or illustrations. Storybooks with scientific errors can be used toward the end of a lesson to teach students how to identify and correct the inaccurate science. For example, as mentioned earlier in this chapter, in the "Amazing Caterpillars" lesson in Chapter 15 students are asked to retell the story of Eric Carle's *The Very Hungry Caterpillar*, a storybook that contains some scientific inaccuracies. Books like this can be powerful vehicles for assessing the ability of learners to analyze the scientific accuracy of a text.

Non-Narrative Information Books

Non-narrative information books are factual texts that introduce a topic, describe the attributes of the topic, or describe typical events that occur. The focus of these texts is on the subject matter, not specific characters. The vocabulary is typically technical. Readers can enter the text at any point in the book. Many contain features found in nonfiction such as a table of contents, bold-print vocabulary words, a glossary, and an index. There is research to suggest that these types of books are "the best resources for fostering children's scientific concepts as well as their appropriation of science discourse" (Pappas 2006). Young children tend to be less familiar with this genre and need many opportunities to experience this type of text. Using non-narrative information books will help students become familiar with the structure of textbooks, as well as "real-world" reading, which is primarily nonfiction. Teachers may want to read only those sections that provide the concepts and facts needed to meet particular science objectives.

One example of non-narrative information writing is the book *Wind Energy: Blown Away!* which contains nonfiction text features such as a table of contents, bold-print words, insets, a glossary, and an index. This book is featured in Chapter 9, "Harnessing the Wind." The appropriate placement of non-narrative information text in a science learning cycle is after students have had the opportunity to explore concepts through hands-on activities. At that point, students are engaged in the topic and are motivated to read the non-narrative informational text to learn more.

Narrative Information Books

Narrative information books, sometimes called hybrid books, provide an engaging format for factual information. They communicate a sequence of factual events over time and sometimes recount the events of a specific case to generalize to all cases. When using these books within science instruction, establish a purpose for reading so that students focus on the science content rather than the storyline. In some cases, teachers may want to read the book one time through for the aesthetic components of the book and a second time for specific science content. *Fossil,* an example of a narrative information text, is used in Chapter 16, "Fossils Tell of Long Ago." This narrative begins with a girl finding a fossil and describes the life of that animal and how it became fossilized. The narrative information genre can be used at any point within a science learning cycle. This genre can be both engaging and informative.

Dual-Purpose Books

Dual-purpose books are intended to serve two purposes: present a story and provide facts. They employ a format that allows readers to use the book like a storybook or to use it like a non-narrative information book. Sometimes information can be found in the running text, but more frequently it appears in insets and diagrams. Readers can enter on any page to access specific facts, or they can read the book through as a story. You can use the story component of a dual-purpose book to engage the reader at the beginning of the science learning cycle. For example, Chapter 14, "Ducks Don't Get Wet," uses the book, *Just Ducks!* to engage the students in learning about duck structures and behaviors. The story appears in a different font size and type than the facts presented in the book, making it easier for the reader to distinguish between the two.

Dual-purpose books typically have little science content within the story. Most of the informational ideas are found in the insets and diagrams, or in the case of *Just Ducks!* in a different font and portion of the page. If the insets and diagrams are read, discussed, explained, and related to the story, these books can be very useful in helping students refine concepts and acquire scientific vocabulary *after* they have had opportunities for hands-on exploration.

Using Fiction and Nonfiction Texts

It can be useful to pair fiction and nonfiction books in read-alouds to round out the science content being presented. Because fiction books tend to be very engaging for students, they can be used to hook students at the beginning of a science lesson. But most of the reading people do in everyday life is nonfiction. We are immersed in informational text every day, and we must be able to comprehend it to be successful in school, at work, and in society. Nonfiction books and other informational text such as articles should be used frequently in the elementary classroom. They often include text structures that differ from stories, and the opportunity to experience these structures in read-alouds can strengthen students' abilities to read and understand informational text. Duke (2004) recommends four strategies to help teachers improve students' comprehension of informational text:

- Increase students' access to informational text.
- Increase the time they spend working with informational text.
- Teach comprehension strategies through direct instruction.
- Create opportunities for students to use informational text for authentic purposes.

Even More Picture-Perfect Science Lessons addresses these recommendations in several ways. The lessons expose students to a variety of nonfiction picture books, articles, and websites on science topics, thereby increasing access to informational text. Various tools (e.g., card sorts, close reading, stop and try it; see Chapter 2 for a complete list of these tools) help enhance students' comprehension of the informational text by increasing the time they spend working with it. Each lesson includes instructions for explicitly teaching comprehension strategies within the learning cycle. The inquiry-based lessons provide an authentic purpose for reading informational text, as students are motivated to read or listen in order to find the answers to questions generated within the inquiry activities.

References

Butzow, J., and C. Butzow. 2000. *Science through children's literature: An integrated approach.* Portsmouth, NH: Teacher Ideas Press.

Casteel, C. P., and B. A. Isom. 1994. Reciprocal processes in science and literacy learning. *The Reading Teacher* 47: 538–544.

Colburn, A. 2003. *The lingo of learning: 88 education terms every science teacher should know*. Arlington, VA: NSTA Press.

Donovan, C., and L. Smolkin. 2002. Considering genre, content, and visual features in the selection of trade books for science instruction. *The Reading Teacher* 55: 502–520.

Duke, N. K. 2004. The case for informational text. *Educational Leadership* 61: 40–44.

Harvey, S., and A. Goudvis. 2000. *Strategies that work: Teaching comprehension to enhance understanding.* York, ME: Stenhouse Publishers.

Kralina, L. 1993. Tricks of the trades: Supplementing your science texts. *The Science Teacher* 60 (9): 33–37.

Martin, D. J. 1997. *Elementary science methods: A constructivist approach.* Albany, NY: Delmar.

Mayer, D. A. 1995. How can we best use children's literature in teaching science concepts? *Science and Children* 32 (6): 16–19, 43.

Miller, K. W., S. F. Steiner, and C. D. Larson. 1996. Strategies for science learning. *Science and Children* 33 (6): 24–27.

Morrow, L. M., M. Pressley, J. K. Smith, and M. Smith. 1997. The effect of a literature-based program integrated into literacy and science instruction with children from diverse backgrounds. *Reading Research Quarterly* 32: 54–76.

National Research Council (NRC). 2012. *A framework for K–12 science education: Practices, crosscutting concepts, and core ideas.* Washington, DC: National Academies Press.

Pappas, C. 2006. The information book genre: Its role in integrated science literacy research and practice. *Reading Research Quarterly* 41 (2): 226–250.

Rice, D. C. 2002. Using trade books in teaching elementary science: Facts and fallacies. *The Reading Teacher* 55 (6): 552–565.

Romance, N. R., and M. R. Vitale. 1992. A curriculum strategy that expands time for in-depth elementary science instruction by using science-based reading strategies: Effects of a year-long study in grade four. *Journal of Research in Science Teaching* 29: 545–554.

Short, K. G., and J. Armstrong. 1993. Moving toward inquiry: Integrating literature into the science curriculum. *New Advocate* 6 (3): 183–200.

Strickland, D. S., and L. M. Morrow, eds. 2000. *Beginning reading and writing*. New York: Teachers College Press.

Tyson, H., and A. Woodward. 1989. Why aren't students learning very much from textbooks? *Educational Leadership* 47 (3): 14–17.

Children's Books Cited

Aliki. 1981. *Digging up dinosaurs*. New York: HarperTrophy.

Carle, E. 1981. *The very hungry caterpillar.* New York: Philomel.

Cleary, B. 1968. *Ramona the pest.* New York: HarperCollins.

Davies, N. 2012. *Just ducks!* Somerville, MA: Candlewick Press.

DeWitt, L. 1993. *What will the weather be?* New York: HarperCollins.

Ewart, C. 2004. *Fossil.* New York: Walker Books for Young Readers.

Hansen, A. 2010. *Wind energy: Blown away!* New York: PowerKids Press.

Henkes, K. 2003. *Wemberly's ice-cream star.* New York: Greenwillow Books.

Hesse, K. 1999. *Come on, rain!* New York: Scholastic.

Lobel, A. 1970. *Frog and Toad Are Friends.* Harper & Row.

Rey, H. A. 1973. *Curious George.* Boston: Houghton Mifflin.

Sendak, M. 1988. *Where the wild things are*. New York: HarperCollins.

Seuss, Dr. 1960. *One fish two fish red fish blue fish.* New York: Random House Books for Young Readers

Silverstein, S. 1974. *Where the sidewalk ends.* New York: HarperCollins.

Simon, S. 1984. *The moon*. Salem, OR: Four Winds.

Stone, J. 2003. *The monster at the end of this book*. New York: Golden Books.

White, E. B. 1952. *Charlotte's web.* New York: HarperCollins.

Reading Aloud

This chapter addresses some of the research supporting the importance of reading aloud, tips to make your read-aloud time more valuable, descriptions of Harvey and Goudvis's six key reading strategies (2000), and tools you can use to enhance students' comprehension during read-aloud time.

Why Read Aloud?

Being read to is the most influential element in building the knowledge required for eventual success in reading (Anderson et al. 1985). It improves reading skills, increases interest in reading and literature, and can even improve overall academic achievement. A good reader demonstrates fluent, expressive reading and models the thinking strategies of proficient readers, helping to build background knowledge and fine-tune students' listening skills. When a teacher does the reading, children's minds are free to anticipate, infer, connect, question, and comprehend (Calkins 2000). In addition, being read to is risk free. In *Yellow Brick Roads: Shared and Guided Paths to Independent Reading 4–12*, Allen (2000) says, "For students who struggle with word-by-word reading, experiencing the whole story can finally give them a sense of the wonder and magic of a book" (p. 45).

Reading aloud is appropriate in all grade levels and for all subjects. Appendix A of the Common Core State Standards for English Language Arts and Literacy (NGA for Best Practices and CCSSO 2010) states that "children in the early grades—particularly kindergarten through grade 3—benefit from participating in rich, structured conversations with an adult in response to written texts that are read aloud, orally comparing and contrasting as well as analyzing and synthesizing" (p. 27). Reading aloud is important not only when children can't read well on their own but also after they have become proficient readers (Anderson et al. 1985). Allen (2000) supports this view: "Given the body of research supporting the importance of read-

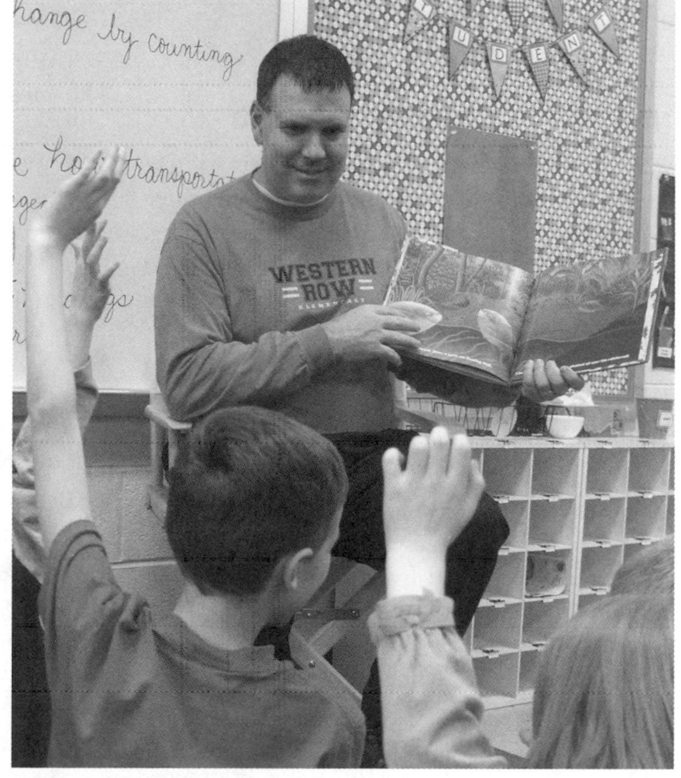

PRINCIPAL MESSER ENJOYS READ-ALOUD TIME.

aloud for modeling fluency, building background knowledge, and developing language acquisition, we should remind ourselves that those same benefits occur when we extend read-aloud beyond the early years" (p. 44). Likewise, the Common Core advocates the use of read-alouds in upper elementary, stating,

> Because children's listening comprehension likely outpaces reading comprehension until the middle school years, it is particularly important that students in the earliest grades build knowledge through being read to as well as through reading, with the balance gradually shifting to reading independently. By reading a story or nonfiction selection aloud, teachers allow children to experience written language without the burden of decoding, granting them access to content that they may not be able to read and understand by themselves. Children are then free to focus their mental energy on the words and ideas presented in the text, and they will eventually be better prepared to tackle rich written content on their own. (NGA for Best Practices and CCSSO 2010, Appendix A, p. 27).

Ten Tips for Reading Aloud

We have provided a list of tips to help you get the most from your read-aloud time. Using these suggestions can help set the stage for learning, improve comprehension of science material, and make the read-aloud experience richer and more meaningful for both you and your students.

1. Preview the Book

Select a book that meets your science objectives and lends itself to reading aloud. Preview it carefully before sharing it with the students. Are there any errors in scientific concepts or misinformation that could be inferred from the text or illustrations? If the book is not in story form, is there any nonessential information you could omit to make the read-aloud experience better? If you are not going to read the whole book, choose appropriate starting and stopping points before reading. Consider generating questions and inferences about the book in advance and placing them on sticky notes inside the book to help you model your thought processes as you read aloud.

2. Set the Stage

Because reading aloud is a performance, you should pay attention to the atmosphere and physical setting of the session. Gather the students in a special reading area, such as on a carpet or in a semicircle of chairs. Seat yourself slightly above them. Do not sit in front of a bright window where the glare will keep students from seeing you well or in an area where students can be easily distracted. You may want to turn off the overhead lights and read by the light of a lamp or use soft music as a way to draw students into the mood of the text. Establish expectations for appropriate behavior during read-aloud time, and, before reading, give the students an opportunity to settle down and focus their attention on the book.

3. Celebrate the Author and Illustrator

Tell students the names of the author and the illustrator before reading. Build connections by asking students if they have read other books by the author or illustrator. Increase interest by sharing facts about the author or illustrator from the book's dust jacket or from library or internet research. The following resources are useful for finding information on authors and illustrators:

Books

Kovacs, D., and J. Preller. 1991. *Meet the authors and illustrators: 60 creators of favorite children's books talk about their work.* Vol. 1. New York: Scholastic.

Kovacs, D., and J. Preller. 1993. *Meet the authors and illustrators: 60 creators of favorite children's books talk about their work.* Vol. 2. New York: Scholastic.

Peacock, S. 2003. *Something about the author: Facts and pictures about authors and illustrators of books for young people.* Vol. 135. Farmington Hills, MI: Gale Group.

Preller, J. 2001. *The big book of picture-book authors and illustrators.* New York: Scholastic.

Website

www.teachingbooks.net—This website continually identifies, catalogs, and maintains reliable links to resources on children's books and their authors and illustrators and organizes theses resources into categories relevant to teachers' needs.

4. Read With Expression

Practice reading aloud to improve your performance. Can you read with more expression to more fully engage your audience? Try louder or softer speech, funny voices, facial expressions, or gestures. Make eye contact with your students every now and then as you read. This strengthens the bond between reader and listener, helps you gauge your audience's response, and cuts down on off-task behaviors. Read slowly enough that your students have time to build mental images of what you are reading, but not so slowly that they lose interest. When reading a nonfiction book aloud, you may want to pause after reading about a key concept to let it sink in and then reread that part. At suspenseful parts in a storybook, use dramatic pauses or slow down and read softly. This can move the audience to the edge of their seats!

5. Share the Pictures

Don't forget the power of visual images to help students connect with and comprehend what you are reading. Make sure that you hold the book in such a way that students can see the pictures on each page. Read captions if appropriate. In some cases, you may want to hide certain pictures so students can visualize what is happening in the text before you reveal the illustrator's interpretation.

6. Encourage Interaction

Keep chart paper and markers nearby in case you want to record questions or new information. Try providing students with "think pads" in the form of sticky notes to write on as you read aloud. Not only does this help extremely active children keep their hands busy while listening, but it also encourages students to interact with the text as they jot down questions or comments. After the read-aloud, have students share their questions and comments. You may want students to place their sticky notes on a class chart whose subject is the topic being studied. Another way to encourage interaction without taking the time for each student to ask questions or comment is to do an occasional "turn and talk" during the read-aloud. Stop reading, ask a question, allow thinking time, and then have each student share ideas with a partner.

7. Keep the Flow

Although you want to encourage interaction during a read-aloud, avoid excessive interruptions that may disrupt fluent, expressive reading. Aim for a balance between allowing students to hear the language of the book uninterrupted and providing them with opportunities to make comments, ask questions, and share connections to the reading. You may want to read the book all the way through one time so students can enjoy the aesthetic components of the story, and then go back and read the book for the purpose of meeting the science objectives.

8. Model Reading Strategies

As you read aloud, it is important that you help children access what they already know and build bridges to new understandings. Think out loud, model your questions for the author, and make connections to yourself, other books, and the world. Show students how to determine the important parts of the text or story, and demonstrate how you synthesize meaning from the text. Modeling these reading comprehension strategies when appropriate before, during, and/or after reading helps students internalize the strategies and begin to use them in their own reading. Six key strategies are described in detail later in this chapter.

9. Don't Put It Away

Keep the read-aloud book accessible to students after you read it. They will want to get a close-up look at the pictures and will enjoy reading the book independently. Don't be afraid of reading the same book more than once—children benefit from the repetition.

10. Have Fun

Let your passion for books show. It is contagious! Read nonfiction books with interest and wonder. Share your thoughts, question the author's intent, synthesize meaning out loud, and voice your own connections to the text. When reading a story, let your emotions show—laugh at the funny parts and cry at the sad parts. Seeing an authentic response from the reader is important for students. If you read with enthusiasm, read-aloud time will become special and enjoyable for everyone involved.

We hope these tips will help you and your students reap the many benefits of read-alouds. As Miller (2002) writes in *Reading With Meaning: Teaching Comprehension in the Primary Grades,* "Learning to read should be a joyful experience. Give children the luxury of listening to well-written stories with interesting plots, singing songs and playing with their words, and exploring a wide range of fiction, nonfiction, poetry and rhymes. … Be genuine. Laugh. Love. Be patient. You're creating a community of readers and thinkers" (p. 26).

Reading Comprehension Strategies

Children's author Madeleine L'Engle (1995) says, "Readers usually grossly underestimate their own importance. If a reader cannot create a book along with the writer, the book will never come to life. The author and the reader … meet on the bridge of words" (p. 34). It is our responsibility as teachers, no matter what subjects we are assigned to teach, to help children realize the importance of their own thoughts and ideas as they read. Modeling our own thinking as we read aloud is the first step. Becoming a proficient reader is an ongoing, complex process, and children need to be explicitly taught the strategies that good readers use. In *Strategies That Work,* Harvey and Goudvis (2000) identify six key reading strategies essential to achieving full understanding when we read. These strategies are used where appropriate in each lesson and are seamlessly embedded into the 5E Model. The strategies should be modeled as you read aloud to students from both fiction and nonfiction texts.

Research shows that explicit teaching of reading comprehension strategies can foster comprehension development (Duke and Pearson 2002). Explicit teaching of the strategies is the initial step in the gradual-release-of-responsibility approach to delivering reading instruction (Fielding and Pearson 1994). During this first phase of the gradual-release method, the teacher *explains* the strategy, demonstrates *how* and *when* to use the strategy, explains *why* it is worth using, and *thinks aloud* to model the mental processes used by good readers. Duke (2004, p. 42) describes the process in this way:

> *I often discuss the strategies in terms of good readers, as in "Good readers think about what might be coming next." I also model the uses of comprehension strategies by thinking aloud as I read. For example, to model the importance of monitoring understanding, I make comments such as, "That doesn't make sense to me because …" or "I didn't understand that last part—I'd better go back.*

Using the teacher-modeling phase within a science learning cycle will reinforce what students do during reading instruction, when the gradual-release-of-responsibility model can be continued. When students have truly mastered a strategy, they are able to apply it to a variety of texts and curricular areas and can explain how the strategy helps them construct meaning.

Descriptions of the six key reading comprehension strategies featured in *Strategies That Work* (Harvey and Goudvis 2000) follow. The icon highlights these strategies here and within the lessons.

Making Connections

Making meaningful connections during reading can improve learners' comprehension and engagement by helping them better relate to what they read. Comprehension breakdown that occurs when reading or listening to expository text can come from a lack of prior information. These three techniques can help readers build background knowledge where little exists:

- *Text-to-self connections* occur when readers and listeners link the text to their past experiences or background knowledge.
- *Text-to-text connections* occur when readers and listeners recognize connections from one book to another.
- *Text-to-world connections* occur when readers and listeners connect the text to events or issues in the real world.

Questioning

Proficient readers ask themselves questions before, during, and after reading. Questioning allows readers to construct meaning, find answers, solve problems, and eliminate confusion as they read. It motivates readers to move forward in the text. Harvey and Goudvis (2000) write, "A reader with no questions might just as well abandon the book. When our students ask questions and search for answers, we know that they are monitoring comprehension and interacting with the text to construct meaning, which is exactly what we hope for in developing readers" (p. 82). Asking questions is not only a critical reading skill but is also at the heart of scientific inquiry and can lead students into meaningful investigations.

Visualizing

Visualizing is the creation of mental images while reading or listening to text. Mental images are created from the learner's emotions and senses, making the text more concrete and memorable. Imagining the sensory qualities of things described in a text can help engage learners and stimulate their interest in the reading. When readers form pictures in their minds, they are also more likely to stick with a challenging text. During a reading, you can stop and ask students to visualize the scene. What sights, sounds, smells, and colors are they imagining?

Inferring

Reading between the lines, or inferring, involves a learner's merging clues from the reading with prior knowledge to draw conclusions and interpret the text. Good readers make inferences before, during, and after reading. Inferential thinking is also an important science skill and can be reinforced during reading instruction.

Determining Importance

Reading to learn requires readers to identify essential information by distinguishing it from nonessential details. Deciding what is important in the text depends on the purpose for reading. In *Even More Picture-Perfect Science Lessons,* each lesson's science objectives determine importance. Learners read or listen to the text to find answers to specific questions, to gain understanding of science concepts, and to identify science misconceptions.

Synthesizing

In synthesizing, readers combine information gained through reading with prior knowledge and experience to form new ideas. To synthesize, readers must stop, think about what they have read, and contemplate its meaning before continuing on through the text. The highest level of synthesis involves those "aha!" moments when readers achieve new insight and, as a result, change their thinking.

Tools to Enhance Comprehension

We have identified several activities and organizers that can enhance students' science understanding and reading comprehension in the lessons. These tools, which support the reading comprehension

strategies from *Strategies That Work* listed above, are briefly described on the following pages and in more detail within the lessons.

Anticipation Guides

Anticipation guides (Herber 1978) are sets of questions that serve as a pre- or postreading activity for a text. They can be used to activate and assess prior knowledge, determine misconceptions, focus thinking on the reading, and motivate reluctant readers by stimulating interest in the topic. An anticipation guide should revolve around four to six key concepts from the reading that learners respond to before reading. They will be motivated to read or listen carefully to find the evidence that supports their predictions. After reading, learners revisit their anticipation guide to check their responses. In a revised extended anticipation guide (Duffelmeyer and Baum 1992), learners are required to justify their responses and explain why their choices were correct or incorrect.

Card Sorts

Card sorts help learners understand the relationships among key concepts and help teach classification. They can also reveal misconceptions and increase motivation to read when used as a prereading activity. Learners are asked to sort words or phrases written on cards into different categories, or sequence the events described on the cards. In an "open sort," learners sort the cards into categories of their own making or sequence events any way they wish. They can re-sort and re-sequence to help refine their understanding of concepts or events. In a "closed sort," the teacher gives them the categories for sorting or provides more information for correctly sequencing their cards.

Close Reading

A *close reading* is a careful and purposeful reading and rereading of a brief but complex text in which the students focus not only on the content but also on critical, high-level reading skills such as author's purpose, point of view, theme, word meaning, and text structures. Close reading is meant to raise the rigor of reading, so it is typically done with a text that is at a slightly higher level than the level at which the student can read independently.

There are many variations on the steps of a close reading. One example is seen in Chapter 9, "Harnessing the Wind." In this lesson students begin a close reading activity by reading an article "cold" with no front-loading of vocabulary or prior discussion of content or structure. Next they write a brief summary of the important points the author is trying to convey and share it with a partner. The teacher then reads the text aloud while the students listen closely. (This practice supports the engagement of all students, especially those who struggle with reading the text independently.) After the teacher reads the text aloud, students reread carefully so that they can answer a series of text-dependent questions. They must refer back to what the text says explicitly and make logical inferences from it. Finally, students discuss the text (rather than their personal reflections) in pairs or in small groups.

When used judiciously and with appropriately complex text,

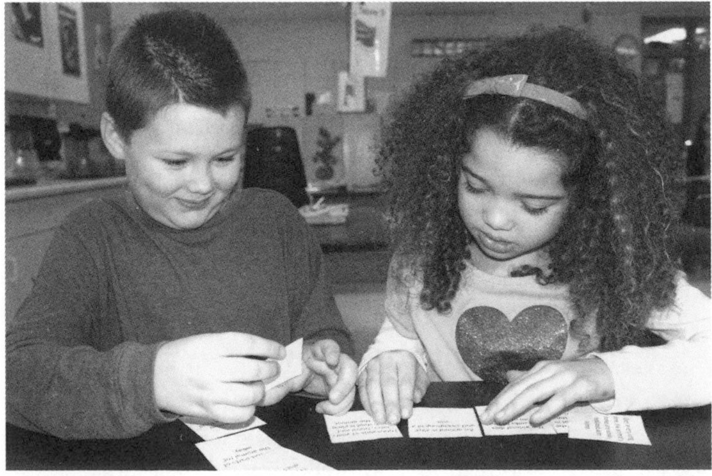

SEQUENCING FOSSIL CARDS IN A CARD SORT

DEMONSTRATING THAT AIR HAS WEIGHT USING THE STOP-AND-TRY-IT FORMAT

close reading can be an effective strategy for deepening content knowledge and learning to read like an expert.

Rereading

Nonfiction text is often full of unfamiliar ideas and difficult vocabulary. *Rereading* content for clarification is an essential skill of proficient readers, and you should model this frequently. Rereading content for a different purpose can aid comprehension. For example, you might read aloud a text for enjoyment and then revisit the text to focus on specific science content.

Sketch to Stretch

During *sketch to stretch* (Seigel 1984), learners pause to reflect on the text and do a comprehension self-assessment by drawing on paper the images they visualize in their heads during reading. They might illustrate an important event from the text, sketch the characters in a story, or make a labeled diagram. Have students use pencils so they understand that the focus should be on collecting their thoughts rather than creating a piece of art. You may want to use a timer so students understand that sketch to stretch is a brief pause to reflect quickly on the reading. Students can share and explain their drawings in small groups after sketching.

Stop and Try It

Stop and try it is a read-aloud format in which the teacher stops reading the text periodically to allow students to observe a demonstration or take part in a hands-on activity to better understand the content being presented. For example, in Chapter 8, "The Wind Blew," we recommend that the teacher stop reading the book *I Face the Wind* at key points to allow students to perform some of the activities described in the book. This way, students have an experience from which to connect the information they are learning from the book.

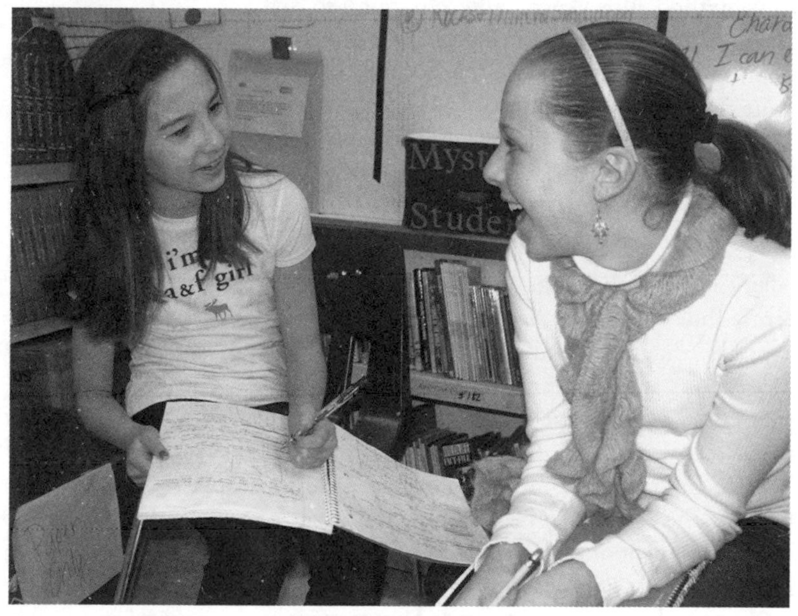

COMPARING SUMMARIES IN A CLOSE READING ACTIVITY

USING THE TABLE OF CONTENTS

Turn and Talk

Learners each pair up with a partner to share their ideas, explain concepts in their own words, or tell about a connection they have to the book. This method allows each child to respond so that everyone in the group is involved as either a talker or a listener. Saying "Take a few minutes to share your thoughts with someone" gives students an opportunity to satisfy their needs to express their own thoughts about the reading.

Using Features of Nonfiction

Many nonfiction books include a table of contents, index, glossary, bold-print words, picture captions, diagrams, and charts that provide valuable information. Because children are generally more used to narrative text, they often skip over these text structures. It is important to model how to interpret the information these features provide the reader. To begin, show the cover of a nonfiction book and read the title and table of contents. Ask students to predict what they'll find in the book. Show students how to use the index in the back of the book to find specific information. Point out other nonfiction text structures as you read and note that these features are unique to nonfiction. Model how nonfiction books can be entered at any point in the text, because they generally don't follow a storyline.

Why Do Picture Books Enhance Comprehension?

Students should be encouraged to read a wide range of print materials, but picture books offer many advantages when teaching reading comprehension strategies. Harvey and Goudvis (2000) not only believe that interest is essential to comprehension but also maintain that, because picture books are extremely effective for building background knowledge and teaching content, instruction in reading comprehension strategies during picture book read-alouds allows students to better access that content. In summary, picture books are invaluable for teaching reading comprehension strategies because they are extraordinarily effective at keeping readers engaged and thinking.

References

Allen, J. 2000. *Yellow brick roads: Shared and guided paths to independent reading 4–12.* Portland, ME: Stenhouse Publishers.

Anderson, R. C., E. H. Heibert, J. Scott, and I. A. G. Wilkinson. 1985. *Becoming a nation of readers: The report of the commission on reading.* Champaign, IL: Center for the Study of Reading; Washington, DC: National Institute of Education.

Calkins, L. M. 2000. *The art of teaching reading.* Boston: Pearson Allyn & Bacon.

Duffelmeyer, F. A., and D. D. Baum. 1992. The extended anticipation guide revisited. *Journal of Reading* 35: 654–656.

Duke, N. K. 2004. The case for informational text. *Educational Leadership* 61: 40–44.

Duke, N. K., and P. D. Pearson. 2002. Effective practices for developing reading comprehension. In *What research has to say about reading instruction,* edited by A. E. Farstrup and S. J. Samuels. Newark, DE: International Reading Association.

Fielding, L., and P. D. Pearson. 1994. Reading comprehension: What works? *Educational Leadership* 51 (5): 62–67.

Harvey, S., and A. Goudvis. 2000. *Strategies that work: Teaching comprehension to enhance understanding.* York, ME: Stenhouse Publishers.

Herber, H. 1978. *Teaching reading in the content areas.* Englewood Cliffs, NJ: Prentice Hall.

L'Engle, M. 1995. *Walking on water: Reflections on faith and art.* New York: North Point Press.

Miller, D. 2002. *Reading with meaning: Teaching comprehension in the primary grades.* Portland, ME: Stenhouse Publishers.

National Governors Association Center (NGA) for Best Practices, and Council of Chief State School Officers (CCSSO). 2010. *Common core state standards for English language arts and literacy.* Washington, DC: National Governors Association for Best Practices, Council of Chief State School.

Seigel, M. 1984. Sketch to stretch. In *Reading, writing, and caring,* edited by O. Cochran. New York: Richard C. Owen.

Chapter 3

Teaching Science Through Inquiry

The word *inquiry* brings many different ideas to mind. For some teachers, it may evoke fears of giving up control in the classroom or spending countless hours preparing lessons. For others, it may imply losing the focus of instructional objectives while students pursue answers to their own questions. And for many, teaching science through inquiry is perceived as intriguing but unrealistic. But inquiry doesn't have to cause anxiety for teachers. Simply stated, inquiry is an approach to learning that involves exploring the world and that leads to asking questions, testing ideas, and making discoveries in the search for understanding. There are many degrees of inquiry, and it may be helpful to start with a variation that emphasizes a teacher-directed approach and then gradually builds to a more student-directed approach. As a basic guide, the National Research Council (2000) identifies five essential features for classroom inquiry, shown in Table 3.1.

Essential Features of Classroom Inquiry

The following descriptions illustrate each of the five essential features of classroom inquiry (NRC 2000) using Chapter 7, "Float Your Boat." Any classroom activity that includes all five of these features is considered to be inquiry.

- *Learners are engaged by scientifically oriented questions.* Students are engaged in the question, Which foil boat design will hold the most pennies?

Table 3.1. Five Essential Features of Classroom Inquiry

1. Learners are engaged by scientifically oriented questions.
2. Learners give priority to evidence, which allows them to develop and evaluate explanations that address scientifically oriented questions.
3. Learners formulate explanations from evidence to address scientifically oriented questions.
4. Learners evaluate their explanations in light of alternative explanations, particularly those reflecting scientific understanding.
5. Learners communicate and justify their proposed explanations.

Adapted from Inquiry and the National Science Education Standards: A Guide for Teaching and Learning *(NRC 2000)*

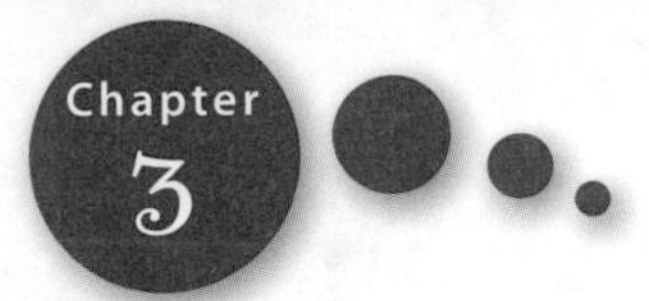

- *Learners give priority to evidence, which allows them to develop and evaluate explanations that address scientifically oriented questions.* Students make boats out of aluminum foil and collect data on how many pennies they hold.
- *Learners formulate explanations from evidence to address scientifically oriented questions.* Students compare their boat designs and results and develop hypotheses about why some boats held more pennies than others.
- *Learners evaluate their explanations in light of alternative explanations, particularly those reflecting scientific understanding.* Students consult a nonfiction book to learn about buoyancy and gravity and the role of these forces in floating and sinking. Then they apply their learning to a new and improved boat design that will hold even more pennies.
- *Learners communicate and justify their proposed explanations.* Students create advertisements for their new boat designs and write persuasive letters, appropriately using the terms *gravity* and *buoyancy* to convince a fictitious person to buy their team's boat.

Inquiry and *A Framework for K–12 Science Education*

Because the term *inquiry* has been interpreted over time in many different ways throughout the science education community, *A Framework for K–12 Science Education* (NRC 2012) attempts to better specify through dimension 1: scientific and engineering practices what is meant by inquiry in science and the range of cognitive, social, and physical practices that it requires. Dimension 1 describes not only the major practices that scientists employ as they investigate and build models and theories about the world, but also a key set of engineering practices that engineers use as they design and build systems. (The term *practices* is used instead of a term such as *skills* to stress that engaging in scientific inquiry requires the use of knowledge and skills simultaneously.) These dimension 1 practices are listed in Table 3.2.

Benefits of Inquiry

Developing an inquiry-based science program is a central tenet of the National Science Education Standards (NRC 1996). So what makes inquiry-based teaching such a valuable method of instruction? Many studies state that it is equal or superior to other instructional modes and results in higher scores on content achievement tests. *Inquiry and the National Science Education Standards* (NRC 2000) summarizes the findings of *How People Learn* (Bransford, Brown, and Cocking 1999), which support the use of inquiry-based teaching. Those findings include the following points:

Table 3.2. Scientific and Engineering Practices for K–12 Science Classrooms

- Asking questions (for science) and defining problems (for engineering)
- Developing and using models
- Planning and carrying out investigations
- Analyzing and interpreting data
- Using mathematics and computational thinking
- Constructing explanations (for science) and designing solutions (for engineering)
- Engaging in argument from evidence
- Obtaining, evaluating, and communicating information

Reprinted from A Framework for K–12 Science Education *(NRC 2012, p. 42).*

- Understanding science is more than knowing facts. Most important is that students understand the major concepts. Inquiry-based teaching focuses on the major concepts, helps students build a strong base of factual information to support the concepts, and gives them opportunities to apply their knowledge effectively.
- Students build new knowledge and understanding on what they already know and believe. Students often hold preconceptions that either are reasonable in only a limited context or are scientifically incorrect. These preconceptions can be resistant to change, particularly when teachers use conventional teaching strategies (Wandersee, Mintzes, and Novak 1994). Inquiry-based teaching uncovers students' prior knowledge and, through concrete explorations, helps them correct misconceptions.
- Students formulate new knowledge by modifying and refining their current concepts and by adding new concepts to what they already know. In an inquiry-based model, students give priority to evidence when they prove or disprove their preconceptions. Their preconceptions are challenged by their observations or the explanations of other students.
- Learning is mediated by the social environment in which learners interact with others. Inquiry provides students with opportunities to interact with others. They explain their ideas to other students and listen critically to the ideas of their classmates. These social interactions require that students clarify their ideas and consider alternative explanations.
- Effective learning requires that students take control of their own learning. When teachers use inquiry, students assume much of the responsibility for their own learning. Students formulate questions, design procedures, develop explanations, and devise ways to share their findings. This makes learning unique and more valuable to each student.
- The ability to apply knowledge to novel situations, that is, transfer of learning, is affected by the degree to which students learn with understanding. Inquiry provides students a variety of opportunities to practice what they have learned and connect to what they already know, and therefore moves them toward application, a sophisticated level of thinking that requires them to solve problems in new situations.

Inquiry learning contributes to better understanding of scientific concepts and skills. In addition, because science inquiry in school is carried out in a social context, inquiry learning contributes to children's social and intellectual development as well (Dyasi 1999). Within an inquiry-based lesson, students work collaboratively to brainstorm questions, design procedures for testing their predictions, carry out investigations, and ask thoughtful questions about other students' conclusions. This mirrors the social context in which "real science" takes place.

What Makes a Good Question?

Questioning lies at the heart of inquiry and is a practice that should be encouraged in any learning setting. As stated in *A Framework for K–12 Science Education* (NRC 2012, p. 54), "Asking questions is essential to developing scientific habits of mind. Even for individuals who do not become scientists or engineers, the ability to ask well-defined questions is an important component of science literacy, helping to make them critical consumers of scientific knowledge."

In an inquiry classroom, curiosity is encouraged, good questions are celebrated, and gathering evidence to answer questions is the focus. According to Inquiry and the National Science Education Standards (NRC 2000, pp. 24–25), "Fruitful inquiries evolve from questions that are meaningful and relevant to students, but they also must be able to be answered by students' observations and scientific knowledge they obtain from reliable sources." A classroom culture that respects and values good questions, offers students opportunities to refine their questions, and incorporates the teaching of effective questioning strategies will promote fruitful inquiries. The teacher in an inquiry classroom plays an important role in helping students identify questions that can lead to interesting and productive investigations, questions that are accessible, manageable, and appropriate to students' developmental level.

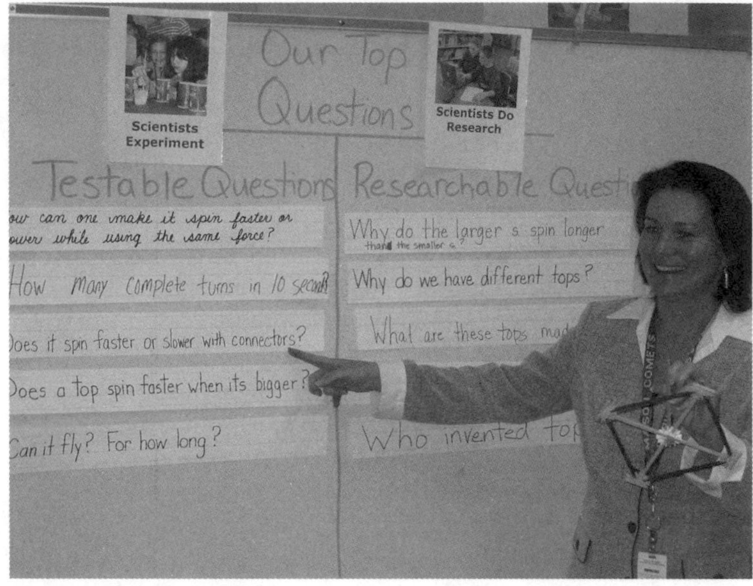

RESULTS OF A QUESTION SORT

Question Sort

One important skill students should develop in science is to understand which questions can be answered by investigation and which cannot. The teacher plays a critical role in guiding the kinds of questions the students pose. Students often ask *why* questions, which cannot be addressed by scientific investigations. For example, "Why does gravity make things fall toward Earth?" is a question that would be impossible to answer in the school setting.

Testable questions, on the other hand, generally begin with *which, can, how can, does,* or *what happens when* and can be investigated using controlled procedures. For example, encouraging students to ask questions such as "How can you slow the fall of an object?" "Which object falls faster, a marble or a basketball?" or "Which materials work best for constructing a toy parachute?" guides them toward investigations that can be done in the classroom.

One way you can help students learn which types of questions are testable and which are not testable is to do a *question sort.* In this activity, you start by providing students with a common experience on a science topic using thought-provoking objects, demonstrations, readings, video clips, and so on. For example, you might provide students with a variety of magnetic toys and objects to explore, do a simple magnet investigation, or watch a video clip about magnets. Next, you ask students to write a question about the topic on a sentence strip or sticky note. Collect all of the sentence strips and read the questions aloud to the class. Explain that the type of investigation a scientist does depends on the questions he or she asks. As a group, you can sort the students' questions into "researchable questions" that can be answered using reliable sources of scientific information and "testable questions" that can be answered by observing, measuring, or doing an experiment. Table 3.3 provides examples of both types of questions.

After taking part in a question sort, students will begin to realize that with a little tweaking, some researchable questions can be turned into testable questions. If students have difficulty coming up with testable questions, you should provide some questions. From your examples, students will gradually begin to learn what testable questions look and sound like.

The next step is to have the class select one of the testable questions and discuss ways to investigate the question. For example, the question "Are larger magnets stronger than smaller magnets?" could be answered by doing a controlled experiment (a "fair test") to collect data on how many paper clips magnets of different sizes can pick up. After investigating the question, students can brainstorm ways to communicate their results,

Table 3.3. Sample Question Sort

Researchable Questions	*Testable Questions*
What makes things magnetic?	Which types of metal objects will stick to a magnet?
What are magnets made of?	Can magnets work through water?
How are magnets used?	How can you make an electromagnet stronger?
What makes some rocks magnetic?	Does a horseshoe magnet have a north and a south pole?
Why do magnets have poles?	What happens to the poles when you break a bar magnet? Will each piece still have a north and south pole?
Why is Earth considered a giant magnet?	Are larger magnets stronger than smaller magnets?

such as with pictures, data tables, graphs, and poster presentations.

Helping students select developmentally appropriate questions is also important. For example, "What will the surface of the Moon look like in a hundred years?" is a question that is scientific but much too complex for elementary students to investigate. A more developmentally appropriate question might be "How does the size of a meteorite affect the size of the crater it makes?" This question can be tested by dropping different-size marbles into a pan of sand, simulating how meteors hit the Moon's surface. It is essential to help students formulate age-appropriate and testable questions to ensure that their investigations are 10.5

The Role of the Teacher

Teaching science through inquiry requires that the teacher take on a different role than the traditional science teacher. "In the inquiry classroom, the teacher's role becomes less involved with direct teaching and more involved with modeling, guiding, facilitating, and continually assessing student work" (Ash and Kluger-Bell 1999, p. 82). One way to guide students and assess their progress as they are engaged in inquiry processes is to ask thoughtful, probing questions. Here are some suggestions for questions to ask students while they are involved in inquiry:

- What would happen if you …?
- What might you try instead?
- What does this remind you of?
- What can you do next time?
- What do you call the things you are using?
- How are you going to do that?
- Is there anything else you could use or do?
- Why did you decide to try that?
- Why do you think that will work?
- Where could you get more information?
- How do you know?
- What is your evidence?

Variations Within Classroom Inquiry

Inquiry-based teaching can vary widely in the amount of guidance and structure you choose to provide. Table 3.4 (p. 24) describes these variations for each of the five essential features of inquiry.

Table 3.4. Inquiry Continuum

	Teacher Guided ⟷ Learner Self-Directed				
	ESSENTIAL FEATURE	VARIATIONS			
1	Learners are engaged by scientifically oriented questions	Learner engages in question provided by teacher or materials	Learner sharpens or clarifies the question provided	Learner selects among questions, poses new questions	Learner poses a question
2	Learners give priority to evidence, which allows them to develop and evaluate explanations that address scientifically oriented questions	Learner is given data and told how to analyze	Learner is given data and asked how to analyze	Learner is directed to collect certain data	Learner determines what constitutes evidence and collects it
3	Learners formulate explanations from evidence to address scientifically oriented questions	Learner is provided with explanations	Learner is given possible ways to use evidence to formulate explanations	Learner is guided in process of formulating explanations from evidence	Learner formulates explanation after summarizing evidence
4	Learners evaluate their explanations in light of alternative explanations, particularly those reflecting scientific understanding	Learner is told connections	Learner is given possible connections	Learner is directed toward areas and sources of scientific knowledge	Learner independently examines other resources and forms the links to explanations
5	Learners communicate and justify their proposed explanations	Learner is given steps and procedures for communication	Learner is provided broad guidelines to sharpen communication	Learner is coached in development of communication	Learner communicates and justifies explanations
	Teacher Guided ⟷ Learner Self-Directed				

Adapted from Inquiry and the National Science Education Standards: A Guide for Teaching and Learning *(NRC 2000).*

The most open form of inquiry takes place in the variations on the right-hand column of the Inquiry Continuum. Most often, students do not have the abilities to begin at that point. For example, students must first learn what makes a question scientifically oriented and testable before they can begin posing such questions themselves. The extent to which you structure what students do determines whether the inquiry is *guided* or *open* inquiry. The more responsibility you take, the more guided the inquiry. The more responsibility the students have, the more open the inquiry. Guided inquiry experiences, such as those on the left-hand side of the Inquiry Continuum, can be effective in focusing learning on the development of particular science concepts. Students, however, must have open inquiry experiences, such as those in the right column of the Inquiry Continuum, to develop the fundamental abilities necessary to do scientific inquiry. Inquiry investigations in the classroom can be highly structured by the teachers so that students proceed toward known outcomes, or inquiry investigations can be free-ranging explorations of unexplained phenomena. Both have their place in science classrooms (NRC 2000).

SIGNALING THE TEACHER WITH A RED/GREEN CUP

One common misconception about inquiry is that all science subject matter should be taught through inquiry. It is not possible or practical to teach all science subject matter through inquiry (NRC 2000). For example, you would not want to teach lab safety through inquiry. Good science teaching requires a variety of approaches and models. *Even More Picture-Perfect Science Lessons* combines guided inquiry investigations with open inquiry investigations. Dunkhase (2000) refers to this approach as "coupled inquiry." The guided inquiries are the lessons presented in each chapter; these lessons generally fall on the left-hand (teacher-guided) side of the Inquiry Continuum. The "Inquiry Place" suggestion box (discussed in depth later in this chapter) at the end of each lesson will produce experiences falling more toward the right-hand, or learner self-directed, side of the Inquiry Continuum.

Checkpoint Labs

One way to manage a guided inquiry is to use a *checkpoint lab.* This type of lab is divided into sections, with a small box located at the end of each section for a teacher check mark or stamp. Each team works at its own pace. A red cup and a green cup with the bottoms taped together is used to signal the teacher. When teams are working, they keep the green cup on top. When teams have a question or when they reach a checkpoint, they signal the teacher by flipping their cups so that red is on top. See Chapter 8, "The Wind Blew," for an example of a checkpoint lab.

Tips for Managing a Checkpoint Lab:

- Give students task cards to assign each student a job.
- Tell students that every member of the team is responsible for recording data and writing re-

sponses. All team members must be at the same checkpoint in order to get the stamp or check mark and continue on to the next section.

- Explain how to use the red/green cups. Tell students that when the green cup is on top, it is a signal to you that the team is progressing with no problems or questions. When the red cup is on top, it is a signal that the team needs the teacher.
- Explain that there are only two situations when students should flip the red cup on top:
    - Everyone on the team is at a checkpoint OR
    - The team has a question.
- Tell students that before they flip the cup to red for a question, they must first ask everyone else on the team the question ("Ask three, then me"). Most of the time, the team will be able to answer the question without asking the teacher.
- When a team reaches a checkpoint and signals you, make sure every member of the team has completed all of the work in that section. Then ask each member a probing question about that part of the lab. Asking each student a question holds them all accountable and allows you to informally assess their learning. Examples of probing questions are

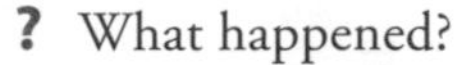

? What happened?
? How do you know?
? What is your evidence?
? Are you surprised by the results? Why or why not?
? What do you think will happen next?

Inquiry Place

An "Inquiry Place" box is provided at the end of each lesson to help you move your students toward more open inquiries. This box lists questions related to the lesson that students may select to investigate. Students may also use the questions as examples to help them generate their own scientifically oriented and testable questions. After selecting one of the questions in the box or formulating their own questions, students can make predictions, design investigations to test their predictions, collect evidence, devise explanations, examine related resources, and communicate their findings.

The "Inquiry Place" boxes suggest that students share the results of their investigations with each other through a poster session. Scientists, engineers, and researchers routinely hold poster sessions to communicate their findings. Here are some suggestions for poster sessions:

- Posters should include a title, the researchers' names, a brief description of the investigation, and a summary of the main findings.
- Observations, data tables, and/or graphs should be included as evidence to justify conclusions.
- The print should be large enough that people can read it from a distance.
- Students should have the opportunity to present their posters to the class.
- The audience in a poster session should examine the evidence, ask thoughtful questions, identify faulty

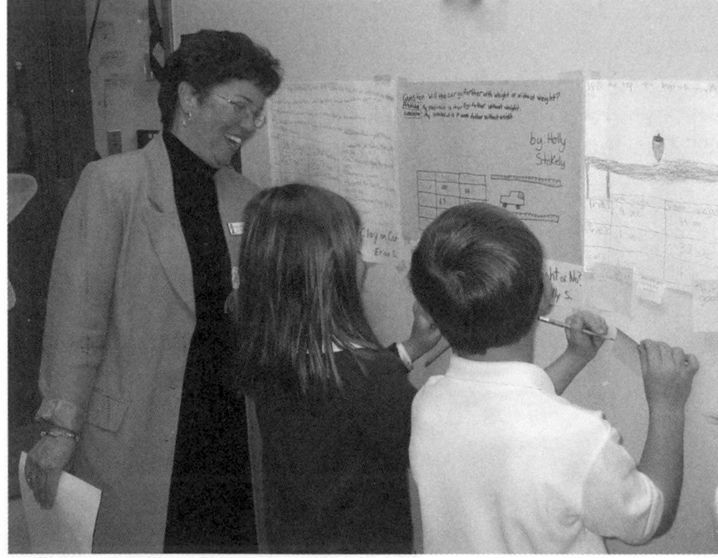

A GALLERY WALK

reasoning, and suggest alternative explanations to presenters in a polite, respectful manner.

Not only do poster sessions mirror the work of real scientists, but they also provide you with excellent opportunities for authentic assessment.

Another way to share students' posters is a gallery walk. In a gallery walk, students put their posters on display for their classmates to view and critique. Students taking the gallery walk use sticky notes to post suggestions, questions, and praise directly on to their classmates' posters. Writing on sticky notes encourages interaction, and the comments provide immediate feedback for the "exhibitors." Here are some guidelines for a gallery walk:

- All necessary information about the investigation should be included on the poster because students will not be giving an oral presentation.
- Like a visit to an art gallery, the gallery walk should be done quietly. Students should be respectful of their classmates' poster displays.
- Students should have the opportunity to read the comments about their own posters and make changes if necessary.

Implementing the guided inquiries in this book along with the Inquiry Place suggestions at the end of each lesson provides a framework for moving from teacher-guided to learner self-directed inquiry. The Inquiry Place Think Sheet (Figure 3.1, p. 28) can help students organize their own inquiries.

References

Ash, D., and B. Kluger-Bell. 1999. Identifying inquiry in the K–5 classroom. In *Inquiry: Thoughts, views, and strategies for the K–5 classroom. Foundations: A monograph for professionals in science, mathematics, and technology education,* Vol. 2. Arlington, VA: National Science Foundation, Division of Elementary, Secondary, and Informal Education in conjunction with the Division of Research, Evaluation, and Communication.

Bransford, J. D., A. L. Brown, and R. Cocking, eds. 1999. *How people learn: Brain, mind, experience, and school.* Washington, DC: National Academies Press.

Dunkhase, J. 2000. Coupled inquiry: An effective strategy for student investigations. Paper presented at the Iowa Science Teachers Section Conference, Des Moines, IA.

Dyasi, H. 1999. What children gain by learning through inquiry. In *Inquiry: Thoughts, views, and strategies for the K–5 classroom. Foundations: A monograph for professionals in science, mathematics, and technology education,* Vol. 2. Arlington, VA: National Science Foundation, Division of Elementary, Secondary, and Informal Education in conjunction with the Division of Research, Evaluation, and Communication.

National Research Council (NRC). 1996. *National science education standards.* Washington, DC: National Academies Press. Also available online at *books.nap.edu/books/0309053269/html/index.html.*

National Research Council (NRC). 2000. *Inquiry and the National Science Education Standards: A guide for teaching and learning.* Washington, DC: National Academies Press. Also available online at *www.nap.edu/books/0309064767/html.*

National Research Council (NRC). 2012. *A framework for K–12 science education: Practices, crosscutting concepts, and core ideas.* Washington, DC: National Academies Press.

National Science Foundation (NSF), Division of Elementary, Secondary, and Information Education in conjunction with the Division of Research, Evaluation, and Communication. 1999. *Inquiry: Thoughts, views, and strategies for the K–5 classroom. Foundations: A monograph for professionals in science, mathematics, and technology education*, Vol. 2. Arlington, VA: National Science Foundation.

Wandersee, J. H., J. J. Mintzes, and J. D. Novak. 1994. Research on alternative conceptions in science. In *Handbook of research on science teaching and learning*, ed. D. L. Gable, 177–210. New York: Macmillan.

Figure 3.1. Inquiry Place Think Sheet

Name: ____

Inquiry Place Think Sheet

1. Topic: ____

2. My questions about the topic: ______________________________

____

____

3. My testable question: ______________________________

____

____

4. Steps I will follow to investigate my question: ______________

____

____

5. Materials I will need: ______________________________

____

____

6. How I will share my findings: ______________________________

____

____

☐ Teacher Checkpoint

BSCS 5E Instructional Model

The guided inquiries in this book are designed using the BSCS 5E Instructional Model, commonly referred to as the 5E Model (or the 5Es). Developed by the Biological Sciences Curriculum Study (BSCS), the 5E Model is a learning cycle based on a constructivist view of learning. Constructivism embraces the idea that learners bring with them preconceived ideas about how the world works. According to the constructivist view, "learners test new ideas against that which they already believe to be true. If the new ideas seem to fit in with their pictures of the world, they have little difficulty learning the ideas … if the new ideas don't seem to fit the learners' picture of reality then they won't seem to make sense. Learners may dismiss them … or eventually accommodate the new ideas and change the way they understand the world" (Colburn 2003, p. 59). The objective of a constructivist model, therefore, is to provide students with experiences that make them reconsider their conceptions. Then, students "redefine, reorganize, elaborate, and change their initial concepts through self-reflection and interaction with their peers and their environment" (Bybee 1997, p. 176). The 5E Model provides a planned sequence of instruction that places students at the center of their learning experiences, encouraging them to explore, construct their own understanding of scientific concepts, and relate those understandings to other concepts. The phases of the 5E model—engage, explore, explain, elaborate, and evaluate—are described here.

Phases of the 5E Model

engage

The purpose of this introductory phase, *engage,* is to capture students' interest. Here you can uncover what students know and think about a topic as well as determine their misconceptions. Engagement activities might include a reading, a demonstration, or other activity that piques students' curiosity.

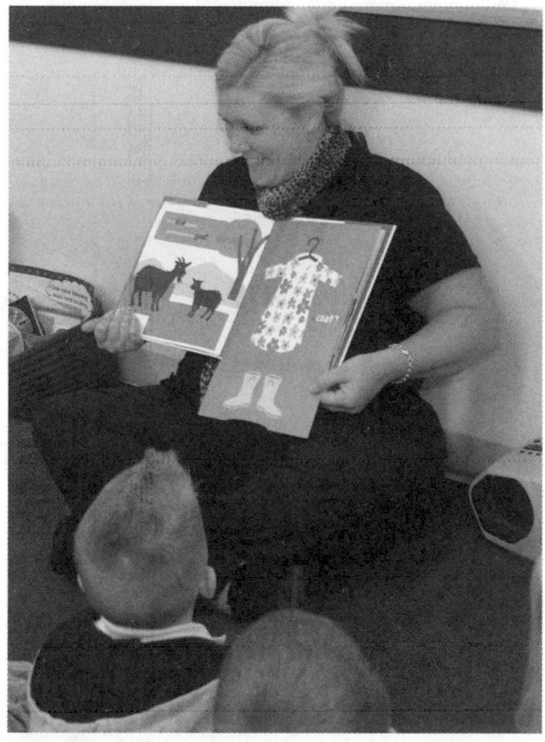

Engaging *students by reading a picture book ("Do You Know Which Ones Will Grow?" Chapter 11)*

explore

In the *explore* phase, you provide students with cooperative exploration activities, giving them common, concrete experiences that help them begin constructing concepts and developing skills. Students can build models, collect data, make and test predictions, or form new predictions. The purpose is to provide hands-on experiences you can use later to formally introduce a concept, process, or skill.

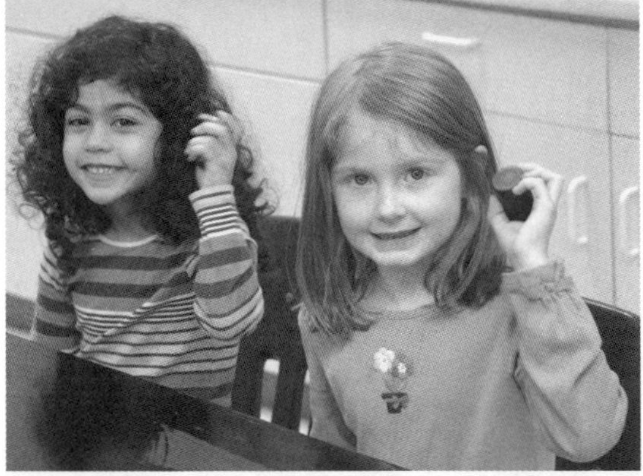

Exploring *sound stations ("Sounds All Around," Chapter 10)*

explain

In the *explain* phase, learners articulate their ideas in their own words and listen critically to one another. You clarify their concepts, correct misconceptions, and introduce scientific terminology. It is important that you clearly connect the students' explanations to experiences they had in the engage and explore phases.

elaborate

At the beginning of the *elaborate* phase, some students may still have misconceptions or may understand the concepts only in the context of the previous exploration. Elaboration activities can help students correct their remaining misconceptions and generalize the concepts in a broader context. These activities also challenge students to apply, extend, or elaborate on concepts and skills in a new situation, resulting in deeper understanding.

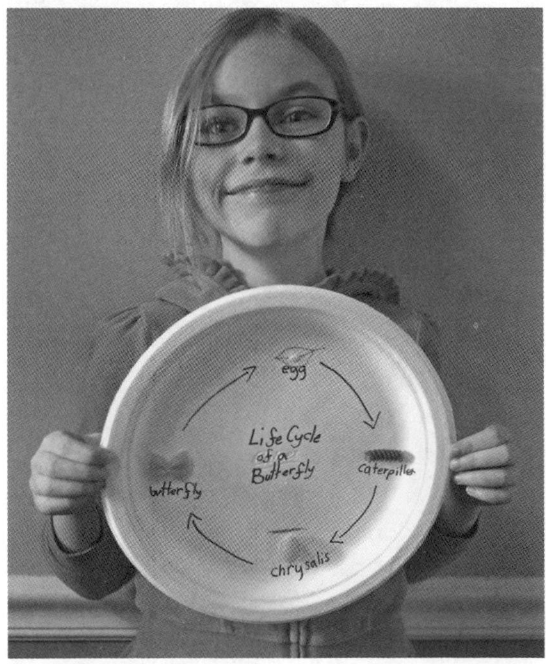

Explaining *the butterfly life cycle with a pasta diagram ("Amazing Caterpillars," Chapter 15)*

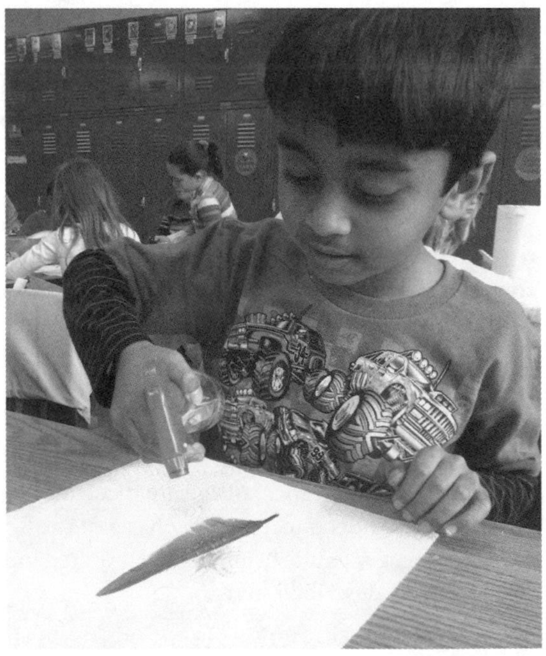

Elaborating *on the learning by simulating how ducks preen ("Ducks Don't Get Wet," Chapter 14)*

evaluate

In the *evaluate* phase, you evaluate students' understanding of concepts and their proficiency with various skills. You can use a variety of formal and informal procedures to assess conceptual understanding and progress toward learning outcomes. The evaluation phase also provides an opportunity for students to test their own understanding and skills.

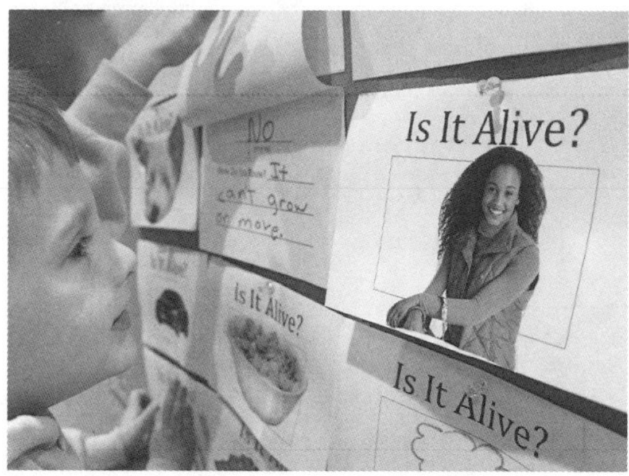

EVALUATING *STUDENT UNDERSTANDING WITH LIFT-THE-FLAP BOOKLETS ("DO YOU KNOW WHICH ONES WILL GROW?" CHAPTER 11)*

Figure 4.1. The BSCS 5Es as a Cycle of Learning

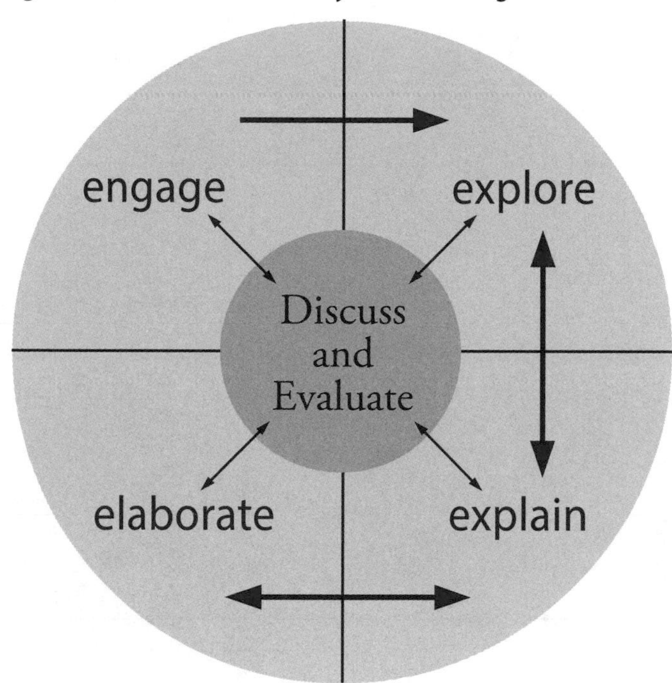

Adapted from Barman, C. R. 1997. The learning cycle revised: A modification of an effective teaching model. *Arlington, VA: Council for Elementary Science International.*

Although the fifth phase is devoted to evaluation, a skillful teacher evaluates throughout the 5E Model, continually checking to see if students need more time or instruction to learn the key points in a lesson. Ways to do this include informal questioning, teacher checkpoints, and class discussions. Each lesson in *Even More Picture-Perfect Science Lessons* also includes a formal evaluation such as extended-response questions or a poster session. These formal evaluations take place at the end of the lesson. A good resource for more information and practical suggestions for evaluating student understanding throughout the 5Es is *Seamless Assessment in Science* by Abell and Volkmann (2006).

Cycle of Learning

In the description above, the 5Es are listed in linear order; however, the model is most effective when you use it as a cycle of learning as in Figure 4.1. Each lesson begins with an engagement activity, but students can reenter the 5E Model at other points in the cycle. For example, in Chapter 14, "Ducks Don't Get Wet," students observe the structures and behaviors of ducks in the explore phase. Then they explain by sharing their observations. Next they elaborate on what they learned about duck preening by experimenting with oil and water. They reenter the explain phase with a nonfiction read-aloud that helps connect the observations they made of the oil and water to the duck preening behavior they observed. Moving from the elaborate phase back into the explain phase gives students the opportunity to add to the knowledge they have constructed thus far in the lesson by learning more about why ducks preen.

The traditional roles of the teacher and student are virtually reversed in the 5E Model. Students

Table 4.1. The BSCS 5Es Teacher

What the Teacher Does		
	CONSISTENT with the BSCS 5E Model	**INCONSISTENT with the BSCS 5E Model**
engage	• Generates interest and curiosity • Raises questions • Assesses current knowledge, including misconceptions	• Explains concepts • Provides definitions and conclusions • Lectures
explore	• Provides time for students to work together • Observes and listens to students as they interact • Asks probing questions to redirect students' investigations when necessary	• Explains how to work through the problem or provides answers • Tells students they are wrong • Gives information or facts that solve the problem
explain	• Asks for evidence and clarification from students • Uses students' previous experiences as a basis for explaining concepts • Encourages students to explain concepts and definitions in their own words, then provides scientific explanations and vocabulary	• Does not solicit the students' explanations • Accepts explanations that have no justification • Introduces unrelated concepts or skills
elaborate	• Expects students to apply scientific concepts, skills, and vocabulary to new situations • Reminds students of alternative explanations • Refers students to alternative explanations	• Provides definite answers • Leads students to step-by-step solutions to new problems • Lectures
evaluate	• Observes and assesses students as they apply new concepts and skills • Allows students to assess their own learning and group process skills • Asks open-ended questions	• Tests vocabulary words and isolated facts • Introduces new ideas or concepts • Promotes open-ended discussion unrelated to the concept

Adapted from Achieving Scientific Literacy: From Purposes to Practices *(Bybee 1997)*

Table 4.2. The BSCS 5Es Student

What the Student Does		
	CONSISTENT with the BSCS 5E Model	**INCONSISTENT with the BSCS 5E Model**
engage	• Asks questions such as "Why did this happen?" "What do I already know about this?" "What can I find out about this? • Shows interest in the topic	• Asks for the "right" answer • Offers the "right" answer • Insists on answers and explanations
explore	• Thinks creatively, but within the limits of the activity • Tests predictions and hypotheses • Records observations and ideas	• Passively allows others to do the thinking and exploring • "Plays around" indiscriminately with no goal in mind • Stops with one solution
explain	• Explains possible solutions to others • Listens critically to explanations of other students and the teacher • Uses recorded observations in explanations	• Proposes explanations from "thin air" with no relationship to previous experiences • Brings up irrelevant experiences and examples • Accepts explanations without justification
elaborate	• Applies new labels, definitions, explanations, and skills in new but similar situation • Uses previous information to ask questions, propose solutions,make decisions, and design experiments • Records observations and explanations	• "Plays around" with no goal in mind • Ignores previous information or evidence • Neglects to record data
evaluate	• Demonstrates an understanding of the concept or skill • Answers open-ended questions by using observations, evidence, and previously accepted explanations • Evaluates his or her own progress and knowledge	• Draws conclusions, not using evidence or previously accepted explanations • Offers only yes-or-no answers and memorized definitions or explanations • Fails to express satisfactory explanations in his or her own words

Adapted from Achieving Scientific Literacy: From Purposes to Practices *(Bybee 1997).*

take on much of the responsibility for learning as they construct knowledge through discovery, whereas in traditional models the teacher is responsible for dispensing information to be learned by the students. Table 4.1 (p. 32) shows actions of the teacher that are consistent with the 5E Model and actions that are inconsistent with the model.

In the 5E Model, the teacher acts as a guide: raising questions, providing opportunities for exploration, asking for evidence to support student explanations, referring students to existing explanations, correcting misconceptions, and coaching students as they apply new concepts. This model differs greatly from the traditional format of lecturing, leading students step-by-step to a solution, providing definite answers, and testing isolated facts. The 5E Model requires the students to take on much of the responsibility for their own learning. Table 4.2 shows the actions of the student that are consistent with the 5E Model and those that are inconsistent with the model.

Using Children's Picture Books in the 5Es

Both fiction and nonfiction picture books can be valuable components of the 5E Model when placed strategically within the cycle. We often begin lessons with a fiction book to pique students' curiosity or motivate them to want to learn more about a science concept. For example, Chapter 7, "Float Your Boat," begins with a story about a boy who makes a toy boat and loses it to the sea for a short time. This read-aloud during the engage phase inspires the question "Why do boats float?" and is followed by activities and a nonfiction read-aloud to find out. A storybook, however, might not be appropriate to use during the explore phase of the 5Es, in which students are participating in concrete, hands-on experiences. Likewise, a storybook might not be appropriate to use during the explain phase to clarify scientific concepts and introduce vocabulary.

You should also avoid using books too early in the learning cycle that contain a lot of scientific terminology or "give away" information students could discover on their own. It is important for students to have opportunities to construct meaning and articulate ideas in their own words before being introduced to scientific vocabulary. Nonfiction books, therefore, are most appropriate to use in the explain phase only after students have had these opportunities. For example, in the explain phase of Chapter 10, "Sounds All Around," students compare the results of their sound station investigations with the information presented in the nonfiction book *Sounds All Around.*

Thoughtful placement of fiction and nonfiction picture books within the BSCS 5E Instructional Model can motivate students to learn about science, allow them to evaluate their findings in light of alternative explanations, and help them understand scientific concepts and vocabulary without taking away from the joy of discovery.

References

Abell, S. K., and M. J. Volkmann. 2006. *Seamless assessment in science: A guide for elementary and middle school teachers.* Chicago: Heinemann; Arlington, VA: NSTA Press.

Bybee, R. W. 1997. *Achieving scientific literacy: From purposes to practices.* Portsmouth, NH: Heinemann.

Colburn, A. 2003. *The lingo of learning: 88 education terms every science teacher should know.* Arlington, VA: NSTA Press.

Connecting to the Standards

A Framework for K–12 Science Education and the Common Core State Standards for English Language Arts

In this book the science and language arts standards that are addressed in the activities for each lesson are clearly identified. On the first page of each chapter, you will find a box titled, "Lesson Objectives Connecting to the *Framework,*" which lists the disciplinary core and component ideas from *A Framework for K–12 Science Education* (NRC 2012) as well as specific grade-level endpoints that the lesson addresses. Throughout the lessons you will find boxes noting the Common Core State Standards for English Language Arts (ELA) that are used during read-alouds and writing assignments. This chapter provides some background information about the *Framework* and the Common Core State Standards for ELA and how our lessons connect to them.

A Framework for K–12 Science Education

A Framework for K–12 Science Education was published by the National Research Council in 2012 and is the first step in a process to create new standards for K–12 science education. The overarching goal of the committee doing this work was "to ensure that by the end of 12th grade, *all* students have some appreciation of the beauty and wonder of science; possess sufficient knowledge of science and engineering to engage in public discussions on related issues; are careful consumers of scientific and technological information related to their everyday lives; are able to continue to learn about science outside school; and have the skills to enter careers of their choice, including (but not limited to) careers in science, engineering, and technology" (NRC 2012, p. 1).

The committee recommended that K–12 science be developed around three major dimensions: (1) scientific and engineering practices, (2) crosscutting concepts, and (3) disciplinary core ideas.

Dimension 1: Scientific and Engineering Practices

This dimension describes eight fundamental practices that scientists use as they investigate and build models and theories about the world, as well as the engineering practices that engineers use as they design and build systems (NRC 2012, p. 42).

1. Asking questions (for science) and defining problems (for engineering)
2. Developing and using models
3. Planning and carrying out investigations
4. Analyzing and interpreting data
5. Using mathematics and computational thinking
6. Constructing explanations (for science) and designing solutions (for engineering)
7. Engaging in argument from evidence
8. Obtaining, evaluating, and communicating information

In *Even More Picture-Perfect Science Lessons* students are engaged in these practices in a guided manner throughout the lessons. At the end of each lesson an "Inquiry Place" box provides suggestions for students to apply these practices in a more student-directed way. For example, in Chapter 8, "The Wind Blew," students learn to implement various scientific practices during a checkpoint lab where they manipulate variables to investigate how force of the wind, direction of the wind, and weight of the boat affect a sailboat's motion. After this guided investigation, they are encouraged to plan and carry out their own investigations about the wind using suggestions in the "Inquiry Place" box.

Some lessons in this book focus more explicitly on the engineering practices. For example, in Chapter 7, "Float Your Boat," students learn how gravity and buoyancy affect boats and then, using engineering practices, they apply what they have learned about these forces to design and build boats. In Chapter 20, "Problem Solvers," students are deeply immersed in the practices of engineers as they design solutions to everyday problems and learn about some of the timeless inventions of Benjamin Franklin. Engineering is sometimes referred to as the "stealth" profession because, although we use many designed objects, we seldom think about the engineering practices involved in the creation and production of these objects. It is important not only to give students an awareness of the work of engineers but also to provide them with opportunities to think like engineers.

Even More Picture-Perfect Science Lessons engages students in these scientific and engineering practices to capture their interest, motivate their continued study, and above all instill in them a sense of wonder about the natural and designed world. The end result is that by actually doing science and engineering rather than merely learning about it, students will recognize that the work of science and engineers is a creative and rewarding endeavor that deeply affects the world they live in.

Dimension 2: Crosscutting Concepts

The seven crosscutting concepts outlined by the *Framework* underlie the lessons in this book:

1. Patterns
2. Cause and effect: mechanism and explanation
3. Scale, proportion, and quantity
4. Systems and system models
5. Energy and matter: flows, cycles, and conservation
6. Structure and function
7. Stability and change

As you implement the lessons in this book, you can use these concepts to help students make connections between disciplines and recognize that the same concept is relevant across different contexts. For example, the crosscutting concept of patterns underlies several lessons. In Chapter 15, "Amazing Caterpillars," students learn that the pattern of egg, larva, pupa, adult is repeated in most insect life cycles. In Chapter 18, "What Will the Weather Be?" students learn how scientists record the patterns of the weather across different times and areas so they can make predictions about what kind of weather will happen next. In Chapter 19, "Sunsets and Shadows," students observe the predictable pattern of the daily changes of the position of the Sun and length and direction of shadows. Noticing patterns is often a first step to organizing and asking scientific questions about why and how the patterns occur. It is important for students to develop ways to recognize, classify, and record patterns in the phenomena they observe, whether it be life cycles, weather, or patterns of change in Earth and sky.

Because students often do not make the connections on their own, it is important for the teacher to make these seven crosscutting concepts explicit for students to help them connect knowledge from different science fields into a coherent and scientifically based view of the world.

Dimension 3: Disciplinary Core Ideas

Disciplinary ideas are grouped in four domains: (1) physical sciences; (2) life sciences; (3) Earth and space sciences; and (4) engineering, technology, and applications of science. Each of these disciplines contains between two and four core ideas, which are then broken down into component ideas. The *Framework* provides grade-level endpoints for grades 2, 5, 8, and 12 for each of the component ideas and, in some cases, boundaries that provide clear guidance regarding expectations for students.

The lesson objectives in this book are closely aligned to a variety of the disciplinary core ideas outlined in dimension 3. At the beginning of each lesson, we provide the discipline, core idea, and component idea as well as the grade-level endpoints we are targeting. Table 5.1 (p. 38) summarizes the correlation between the lessons presented in *Even More Picture-Perfect Science Lessons* and the disciplinary core and component ideas from *A Framework for K–12 Science Education.*

Common Core State Standards for English Language Arts

The Common Core State Standards Initiative (*www.corestandards.org*) is a state-led effort to define the knowledge and skills students should acquire in their K–12 mathematics and ELA courses. It is a result of an extended, broad-based effort to fulfill the charge issued by the states to craft the next generation of K–12 standards to ensure that all students are college and career ready in literacy by the end of high school. The standards are research and evidence based, aligned with college and work expectations, rigorous, and internationally benchmarked. The Common Core suggests that the ELA standards be taught in the context of history/social studies, science, and technical subjects (NGA for Best Practices and CCSO 2010). Grade-specific K–12 standards in reading, writing, speaking, listening, and language are included. Many of these grade-specific standards are used in *Even More Picture-Perfect Science Lessons* through the use of high-quality children's fiction and nonfiction picture books, research-based reading strategies, poster presentations, vocabulary development activities, and various writing assignments. In the boxes titled "Connecting to the Common Core" you will find the Common Core ELA strand(s) and topic the activity addresses, the grade level(s), and the standard number(s) (see Figure 5.1).

Because the codes for the Common Core ELA Standards are listed in the lessons instead of the actual standards statements, we have included the Common Core ELA grade level statements in Table 5.2. This table is not a complete version of the Common Core ELA standards. Rather it includes only the standards we address in our lessons for grades K–5: Reading (literature and informational text), writing, speaking and listening, and language. You can access the complete Common Core State Standards documents for ELA at *www.corestandards.org/ELA-Literacy.*

Figure 5.1. Sample Common Core Box

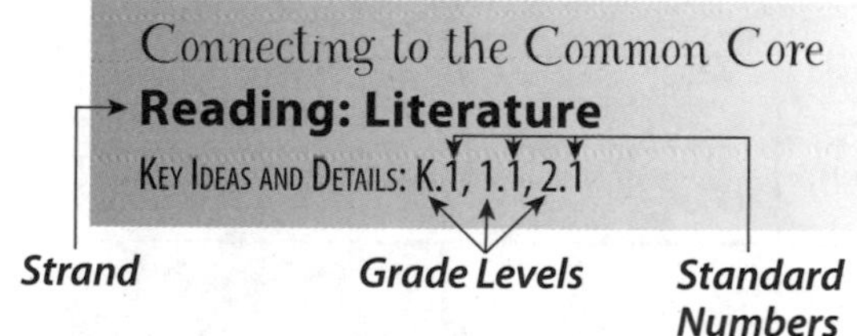

References

National Governors Association Center (NGA) for Best Practices, and Council of Chief State School Officers (CCSSO). 2010. *Common core state standards for English language arts and literacy.* Washington, DC: National Governors Association for Best Practices, Council of Chief State School.

National Research Council. 2012. *A framework for K–12 science education: Practices, crosscutting concepts, and core ideas.* Washington DC: National Academies Press.

Table 5.1. Connecting to A Framework for K–12 Science Education

Lesson	Physical Sciences	Life Sciences	Earth and Space Sciences	Engineering, Technology, and Applications of Science
Chapter 6: Freezing and Melting	K–2 PS1.A Structure and Properties of Matter PS1.B Chemical Reactions			
Chapter 7: Float Your Boat	3–5 PS2.A Forces and Motion PS2.B Types of Interactions			3–5 ETS1.C Optimizing the Design Solution
Chapter 8: The Wind Blew	3–5 PS1.A Structure and Properties of Matter PS3.C Relationship Between Energy and Forces			
Chapter 9: Harnessing the Wind	3–5 PS3.A Definitions of Energy PS3.B Conservation of Energy and Energy Transfer		3–5 ESS3.A Natural Resources	
Chapter 10: Sounds All Around	K–2 PS4.A Wave Properties			
Chapter 11: Do You Know Which Ones Will Grow?		K–2 LS1.C Organization for Matter and Energy Flow in Organisms LS4.D Biodiversity and Humans		
Chapter 12: Seeds on the Move		K–2 LS1.A Structure and Function LS2.A Interdependent Relationships in Ecosystems		

Lesson	Physical Sciences	Life Sciences	Earth and Space Sciences	Engineering, Technology, and Applications of Science
Chapter 13: Unbeatable Beaks		K–2 LS1.A Structure and Function		
Chapter 14: Ducks Don't Get Wet		3–5 LS1.A Structure and Function		
Chapter 15: Amazing Caterpillars		K–2 LS1.B Growth and Development of Organisms		
Chapter 16: Fossils Tell of Long Ago		3–5 LS4.A Evidence of Common Ancestry and Diversity		
Chapter 17: Reduce, Reuse, Recycle!			K–2 ESS3.C Human Impacts on Earth Systems	
Chapter 18: What Will the Weather Be?			3–5 ESS2.D Weather and Climate	
Chapter 19: Sunsets and Shadows			3–5 ESS1.B Earth and the Solar System	
Chapter 20: Problem Solvers				3–5 ETS1.A Defining and Delimiting an Engineering Problem ETS1.B Developing Possible Solutions ETS1.C Optimizing the Design Solution ETS2.B Influence of Engineering, Technology, and Science on Society and the Natural World

Table 5.2 Common Core State Standards for English Language Arts & Literacy in History/Social Studies, Science, and Technical Subjects

Reading Standards for Literature K–2

Kindergartners:	Grade 1 Students:	Grade 2 Students:
	Key Ideas and Details	
1. With prompting and support, ask and answer questions about key details in a text.	1. Ask and answer questions about key details in a text.	1. Ask and answer such questions as *who, what, where, when, why*, and *how* to demonstrate understanding of key details in a text.
2. With prompting and support, retell familiar stories, including key details.	2. Retell stories, including key details, and demonstrate understanding of their central message or lesson.	2. Recount stories, including fables and folktales from diverse cultures, and determine their central message, lesson, or moral.
3. With prompting and support, identify characters, settings, and major events in a story.	3. Describe characters, settings, and major events in a story, using key details.	3. Describe how characters in a story respond to major events and challenges.
	Craft and Structure	
4. Ask and answer questions about unknown words in a text.	4. Identify words and phrases in stories or poems that suggest feelings or appeal to the senses.	4. Describe how words and phrases (e.g., regular beats, alliteration, rhymes, repeated lines) supply rhythm and meaning in a story, poem, or song.
5. Recognize common types of texts (e.g., storybooks, poems).	5. Explain major differences between books that tell stories and books that give information, drawing on a wide reading of a range of text types.	5. Describe the overall structure of a story, including describing how the beginning introduces the story and the ending concludes the action.
6. With prompting and support, name the author and illustrator of a story and define the role of each in telling the story.	6. Identify who is telling the story at various points in a text.	6. Acknowledge differences in the points of view of characters, including by speaking in a different voice for each character when reading dialogue aloud.
	Integration of Knowledge and Ideas	
7. With prompting and support, describe the relationship between illustrations and the story in which they appear (e.g., what moment in a story an illustration depicts).	7. Use illustrations and details in a story to describe its characters, setting, or events.	7. Use information gained from the illustrations and words in a print or digital text to demonstrate understanding of its characters, setting, or plot.
8. (Not applicable to literature)	8. (Not applicable to literature)	8. (Not applicable to literature)
9. With prompting and support, compare and contrast the adventures and experiences of characters in familiar stories.	9. Compare and contrast the adventures and experiences of characters in stories.	9. Compare and contrast two or more versions of the same story (e.g., Cinderella stories) by different authors or from different cultures.
	Range of Reading and Level of Text Complexity	
10. Actively engage in group reading activities with purpose and understanding.	10. With prompting and support, read prose and poetry of appropriate complexity for grade 1.	10. By the end of the year, read and comprehend literature, including stories and poetry, in the grades 2–3 text complexity band proficiently, with scaffolding as needed at the high end of the range.

Reading Standards for Literature 3–5

Grade 3 Students:	Grade 4 Students:	Grade 5 Students:
	Key Ideas and Details	
1. Ask and answer questions to demonstrate understanding of a text, referring explicitly to the text as the basis for the answers.	1. Refer to details and examples in a text when explaining what the text says explicitly and when drawing inferences from the text.	1. Quote accurately from a text when explaining what the text says explicitly and when drawing inferences from the text.
2. Recount stories, including fables, folktales, and myths from diverse cultures; determine the central message, lesson, or moral and explain how it is conveyed through key details in the text.	2. Determine a theme of a story, drama, or poem from details in the text; summarize the text.	2. Determine a theme of a story, drama, or poem from details in the text, including how characters in a story or drama respond to challenges or how the speaker in a poem reflects upon a topic; summarize the text.
3. Describe characters in a story (e.g., their traits, motivations, or feelings) and explain how their actions contribute to the sequence of events.	3. Describe in depth a character, setting, or event in a story or drama, drawing on specific details in the text (e.g., a character's thoughts, words, or actions).	3. Compare and contrast two or more characters, settings, or events in a story or drama, drawing on specific details in the text (e.g., how characters interact).
	Craft and Structure	
4. Determine the meaning of words and phrases as they are used in a text, distinguishing literal from nonliteral language.	4. Determine the meaning of words and phrases as they are used in a text, including those that allude to significant characters found in mythology (e.g., Herculean).	4. Determine the meaning of words and phrases as they are used in a text, including figurative language such as metaphors and similes.
5. Refer to parts of stories, dramas, and poems when writing or speaking about a text, using terms such as chapter, scene, and stanza; describe how each successive part builds on earlier sections.	5. Explain major differences between poems, drama, and prose, and refer to the structural elements of poems (e.g., verse, rhythm, meter) and drama (e.g., casts of characters, settings, descriptions, dialogue, stage directions) when writing or speaking about a text.	5. Explain how a series of chapters, scenes, or stanzas fits together to provide the overall structure of a particular story, drama, or poem.
6. Distinguish their own point of view from that of the narrator or those of the characters.	6. Compare and contrast the point of view from which different stories are narrated, including the difference between first- and third-person narrations.	6. Describe how a narrator's or speaker's point of view influences how events are described.
	Integration of Knowledge and Ideas	
7. Explain how specific aspects of a text's illustrations contribute to what is conveyed by the words in a story (e.g., create mood, emphasize aspects of a character or setting).	7. Make connections between the text of a story or drama and a visual or oral presentation of the text, identifying where each version reflects specific descriptions and directions in the text.	7. Analyze how visual and multimedia elements contribute to the meaning, tone, or beauty of a text (e.g., graphic novel, multimedia presentation of fiction, folktale, myth, poem).
8. (Not applicable to literature)	8. (Not applicable to literature)	8. (Not applicable to literature)
9. Compare and contrast the themes, settings, and plots of stories written by the same author about the same or similar characters (e.g., in books in from a series).	9. Compare and contrast the treatment of similar themes and topics (e.g., opposition of good and evil) and patterns of events (e.g., the quest) in stories, myths, and traditional literature from different cultures.	9. Compare and contrast stories in the same genre (e.g., mysteries and adventure stories) on their approaches to similar themes and topics.
	Range of Reading and Level of Text Complexity	
10. By the end of the year, read and comprehend literature, including stories, dramas, and poetry, at the high end of the grades 2–3 text complexity band independently and proficiently.	10. By the end of the year, read and comprehend literature, including stories, dramas, and poetry, in the grades 4–5 text complexity band proficiently, with scaffolding as needed at the high end of the range.	10. By the end of the year, read and comprehend literature, including stories, dramas, and poetry, at the high end of the grades 4–5 text complexity band independently and proficiently.

Reading Standards for Informational Text K–2

Kindergartners:	Grade 1 Students:	Grade 2 Students:
	Key Ideas and Details	
1. With prompting and support, ask and answer questions about key details in a text.	1. Ask and answer questions about key details in a text.	1. Ask and answer such questions as who, *what, where, when, why,* and *how* to demonstrate understanding of key details in a text.
2. With prompting and support, identify the main topic and retell key details of a text.	2. Identify the main topic and retell key details of a text.	2. Identify the main topic of a multiparagraph text as well as the focus of specific paragraphs within the text.
3. With prompting and support, describe the connection between two individuals, events, ideas, or pieces of information in a text.	3. Describe the connection between two individuals, events, ideas, or pieces of information in a text.	3. Describe the connection between a series of historical events, scientific ideas or concepts, or steps in technical procedures in a text.
	Craft and Structure	
4. With prompting and support, ask and answer questions about unknown words in a text.	4. Ask and answer questions to help determine or clarify the meaning of words and phrases in a text.	4. Determine the meaning of words and phrases in a text relevant to a *grade 2 topic or subject area.*
5. Identify the front cover, back cover, and title page of a book.	5. Know and use various text features (e.g., headings, tables of contents, glossaries, electronic menus, icons) to locate key facts or information in a text.	5. Know and use various text features (e.g., captions, bold print, subheadings, glossaries, indexes, electronic menus, icons) to locate key facts or information in a text efficiently.
6. Name the author and illustrator of a text and define the role of each in presenting the ideas or information in a text.	6. Distinguish between information provided by pictures or other illustrations and information provided by the words in a text.	6. Identify the main purpose of a text, including what the author wants to answer, explain, or describe.
	Integration of Knowledge and Ideas	
7. With prompting and support, describe the relationship between illustrations and the text in which they appear (e.g., what person, place, thing, or idea in the text an illustration depicts).	7. Use the illustrations and details in a text to describe its key ideas.	7. Explain how specific images (e.g., a diagram showing how a machine works) contribute to and clarify a text.
8. With prompting and support, identify the reasons an author gives to support points in a text.	8. Identify the reasons an author gives to support points in a text.	8. Describe how reasons support specific points the author makes in a text.
9. With prompting and support, identify basic similarities in and differences between two texts on the same topic (e.g., in illustrations, descriptions, or procedures).	9. Identify basic similarities in and differences between two texts on the same topic (e.g., in illustrations, descriptions, or procedures).	9. Compare and contrast the most important points presented by two texts on the same topic.
	Range of Reading and Level of Text Complexity	
9. Actively engage in group reading activities with purpose and understanding.	9. With prompting and support, read informational texts appropriately complex for grade 1.	9. By the end of year, read and comprehend informational texts, including history/ social studies, science, and technical texts, in the grades 2–3 text complexity band proficiently, with scaffolding as needed at the high end of the range.
10. Actively engage in group reading activities with purpose and understanding.	10. With prompting and support, read informational texts appropriately complex for grade 1.	10. By the end of year, read and comprehend informational texts, including history/ social studies, science, and technical texts, in the grades 2–3 text complexity band proficiently, with scaffolding as needed at the high end of the range.

Reading Standards for Informational Text 3–5

Grade 3 Students:	Grade 4 Students:	Grade 5 Students:
	Key Ideas and Details	
1. Ask and answer questions to demonstrate understanding of a text, referring explicitly to the text as the basis for the answers.	**1.** Refer to details and examples in a text when explaining what the text says explicitly and when drawing inferences from the text.	**1.** Quote accurately from a text when explaining what the text says explicitly and when drawing inferences from the text.
2. Determine the main idea of a text; recount the key details and explain how they support the main idea.	**2.** Determine the main idea of a text and explain how it is supported by key details; summarize the text.	**2.** Determine two or more main ideas of a text and explain how they are supported by key details; summarize the text.
3. Describe the relationship between a series of historical events, scientific ideas or concepts, or steps in technical procedures in a text, using language that pertains to time, sequence, and cause/effect.	**3.** Explain events, procedures, ideas, or concepts in a historical, scientific, or technical text, including what happened and why, based on specific information in the text.	**3.** Explain the relationships or interactions between two or more individuals, events, ideas, or concepts in a historical, scientific, or technical text based on specific information in the text.
	Craft and Structure	
4. Determine the meaning of general academic and domain-specific words and phrases in a text relevant to a *grade 3 topic or subject area.*	**4.** Determine the meaning of general academic and domain-specific words or phrases in a text relevant to a *grade 4 topic or subject area.*	**4.** Determine the meaning of general academic and domain-specific words and phrases in a text relevant to a *grade 5 topic or subject area.*
5. Use text features and search tools (e.g., key words, sidebars, hyperlinks) to locate information relevant to a given topic efficiently.	**5.** Describe the overall structure (e.g., chronology, comparison, cause/effect, problem/solution) of events, ideas, concepts, or information in a text or part of a text.	**5.** Compare and contrast the overall structure (e.g., chronology, comparison, cause/effect, problem/solution) of events, ideas, concepts, or information in two or more texts.
6. Distinguish their own point of view from that of the author of a text.	**6.** Compare and contrast a firsthand and secondhand account of the same event or topic; describe the differences in focus and the information provided.	**6.** Analyze multiple accounts of the same event or topic, noting important similarities and differences in the point of view they represent.
	Integration of Knowledge and Ideas	
7. Use information gained from illustrations (e.g., maps, photographs) and the words in a text to demonstrate understanding of the text (e.g., where, when, why, and how key events occur).	**7.** Interpret information presented visually, orally, or quantitatively (e.g., in charts, graphs, diagrams, time lines, animations, or interactive elements on Web pages) and explain how the information contributes to an understanding of the text in which it appears.	**7.** Draw on information from multiple print or digital sources, demonstrating the ability to locate an answer to a question quickly or to solve a problem efficiently.
8. Describe the logical connection between particular sentences and paragraphs in a text (e.g., comparison, cause/effect, first/second/third in a sequence).	**8.** Explain how an author uses reasons and evidence to support particular points in a text.	**8.** Explain how an author uses reasons and evidence to support particular points in a text, identifying which reasons and evidence support which point(s).
9. Compare and contrast the most important points and key details presented in two texts on the same topic.	**9.** Integrate information from two texts on the same topic in order to write or speak about the subject knowledgeably.	**9.** Integrate information from several texts on the same topic in order to write or speak about the subject knowledgeably.
	Range of Reading and Level of Text Complexity	
10. By the end of the year, read and comprehend informational texts, including history/social studies, science, and technical texts, at the high end of the grades 2–3 text complexity band independently and proficiently.	**10.** By the end of year, read and comprehend informational texts, including history/social studies, science, and technical texts, in the grades 4–5 text complexity band proficiently, with scaffolding as needed at the high end of the range.	**10.** By the end of the year, read and comprehend informational texts, including history/social studies, science, and technical texts, at the high end of the grades 4–5 text complexity band independently and proficiently.

Writing Standards K–2

Kindergartners:	Grade 1 Students:	Grade 2 Students:
	Text Types and Purposes	
1. Use a combination of drawing, dictating, and writing to compose opinion pieces in which they tell a reader the topic or the name of the book they are writing about and state an opinion or preference about the topic or book (e.g., *My favorite book is …*).	1. Write opinion pieces in which they introduce the topic or name the book they are writing about, state an opinion, supply a reason for the opinion, and provide some sense of closure.	1. Write opinion pieces in which they introduce the topic or book they are writing about, state an opinion, supply reasons that support the opinion, use linking words (e.g., *because, and, also*) to connect opinion and reasons, and provide a concluding statement or section.
2. Use a combination of drawing, dictating, and writing to compose informative/explanatory texts in which they name what they are writing about and supply some information about the topic.	2. Write informative/explanatory texts in which they name a topic, supply some facts about the topic, and provide some sense of closure.	2. Write informative/explanatory texts in which they introduce a topic, use facts and definitions to develop points, and provide a concluding statement or section.
3. Use a combination of drawing, dictating, and writing to narrate a single event or several loosely linked events, tell about the events in the order in which they occurred, and provide a reaction to what happened.	3. Write narratives in which they recount two or more appropriately sequenced events, include some details regarding what happened, use temporal words to signal event order, and provide some sense of closure.	3. Write narratives in which they recount a well-elaborated event or short sequence of events, include details to describe actions, thoughts, and feelings, use temporal words to signal event order, and provide a sense of closure.
	Production and Distribution of Wrting	
4. (Begins in grade 3)	4. (Begins in grade 3)	4. (Begins in grade 3)
5. With guidance and support from adults, respond to questions and suggestions from peers and add details to strengthen writing as needed.	5. With guidance and support from adults, focus on a topic, respond to questions and suggestions from peers, and add details to strengthen writing as needed.	5. With guidance and support from adults and peers, focus on a topic and strengthen writing as needed by revising and editing.
6. With guidance and support from adults, explore a variety of digital tools to produce and publish writing, including in collaboration with peers.	6. With guidance and support from adults, use a variety of digital tools to produce and publish writing, including in collaboration with peers.	6. With guidance and support from adults, use a variety of digital tools to produce and publish writing, including in collaboration with peers.
	Research to Build and Present Knowledge	
7. Participate in shared research and writing projects (e.g., explore a number of books by a favorite author and express opinions about them).	7. Participate in shared research and writing projects (e.g., explore a number of "how-to" books on a given topic and use them to write a sequence of instructions).	7. Participate in shared research and writing projects (e.g., read a number of books on a single topic to produce a report; record science observations).
8. With guidance and support from adults, recall information from experiences or gather information from provided sources to answer a question.	8. With guidance and support from adults, recall information from experiences or gather information from provided sources to answer a question.	8. Recall information from experiences or gather information from provided sources to answer a question.
9. (Begins in grade 4)	9. (Begins in grade 4)	9. (Begins in grade 4)
	Range of Writing	
10. (Begins in grade 3)	10. (Begins in grade 3)	10. (Begins in grade 3)

Writing Standards 3–5

Grade 3 Students:	Grade 4 Students:	Grade 5 Students:
	Text Types and Purposes	
1. Write opinion pieces on topics or texts, supporting a point of view with reasons. **a.** Introduce the topic or text they are writing about, state an opinion, and create an organizational structure that lists reasons. **b.** Provide reasons that support the opinion. **c.** Use linking words and phrases (e.g., *because, therefore, since, for example*) to connect opinion and reasons. **d.** Provide a concluding statement or section.	**1.** Write opinion pieces on topics or texts, supporting a point of view with reasons and information. **a.** Introduce a topic or text clearly, state an opinion, and create an organizational structure in which related ideas are grouped to support the writer's purpose. **b.** Provide reasons that are supported by facts and details. Know final -e and common vowel team conventions for representing long vowel sounds. **c.** Link opinion and reasons using words and phrases (e.g., *for instance, in order to, in addition*). Decode two-syllable words following basic patterns by breaking the words into syllables. **d.** Provide a concluding statement or section related to the opinion presented.	**1.** Write opinion pieces on topics or texts, supporting a point of view with reasons and information. **a.** Introduce a topic or text clearly, state an opinion, and create an organizational structure in which ideas are logically grouped to support the writer's purpose. **b.** Provide logically ordered reasons that are supported by facts and details. **c.** Link opinion and reasons using words, phrases, and clauses (e.g., *consequently, specifically*). **d.** Provide a concluding statement or section related to the opinion presented.
2. Write informative/explanatory texts to examine a topic and convey ideas and information clearly. **a.** Introduce a topic and group related information together; include illustrations when useful to aiding comprehension. **b.** Develop the topic with facts, definitions, and details. **c.** Use linking words and phrases (e.g., *also, another, and, more, but*) to connect ideas within categories of information. **d.** Provide a concluding statement or section.	**2.** Write informative/explanatory texts to examine a topic and convey ideas and information clearly. **a.** Introduce a topic clearly and group related information in paragraphs and sections; include formatting (e.g., headings), illustrations, and multimedia when useful to aiding comprehension. **b.** Develop the topic with facts, definitions, concrete details, quotations, or other information and examples related to the topic. Use context to confirm or self-correct word recognition and understanding, rereading as necessary. **c.** Link ideas within categories of information using words and phrases (e.g., *another, for example, also, because*). **d.** Use precise language and domain-specific vocabulary to inform about or explain the topic.	**2.** Write informative/explanatory texts to examine a topic and convey ideas and information clearly. **a.** Introduce a topic clearly and group related information in paragraphs and sections; include formatting (e.g., headings), illustrations, and multimedia when useful to aiding comprehension. Read grade-level text orally with accuracy, appropriate rate, and expression on successive readings. **b.** Develop the topic with facts, definitions, concrete details, quotations, or other information and examples related to the topic. **c.** Link ideas within categories of information using words and phrases (e.g., *another, for example, also, because*). **d.** Use precise language and domain-specific vocabulary to inform about or explain the topic. **e.** Provide a concluding statement or section related to the information or explanation presented.
3. Write narratives to develop real or imagined experiences or events using effective technique, descriptive details, and clear event sequences. **a.** Establish a situation and introduce a narrator and/or characters; organize an event sequence that unfolds naturally. **b.** Use dialogue and descriptions of actions, thoughts, and feelings to develop experiences and events or show the response of characters to situations. **c.** Use temporal words and phrases to signal event order. **d.** Provide a sense of closure.	**3.** Write narratives to develop real or imagined experiences or events using effective technique, descriptive details, and clear event sequences. **a.** Orient the reader by establishing a situation and introducing a narrator and/or characters; organize an event sequence that unfolds naturally. **b.** Use dialogue and description to develop experiences and events or show the responses of characters to situations. **c.** Use a variety of transitional words and phrases to manage the sequence of events. **d.** Use concrete words and phrases and sensory details to convey experiences and events precisely. **e.** Provide a conclusion that follows from the narrated experiences or events.	**3.** Write narratives to develop real or imagined experiences or events using effective technique, descriptive details, and clear event sequences. **a.** Orient the reader by establishing a situation and introducing a narrator and/or characters; organize an event sequence that unfolds naturally. **b.** Use narrative techniques, such as dialogue, description, and pacing, to develop experiences and events or show the responses of characters to situations. **c.** Use a variety of transitional words, phrases, and clauses to manage the sequence of events. **d.** Use concrete words and phrases and sensory details to convey experiences and events precisely. **e.** Provide a conclusion that follows from the narrated experiences or events.

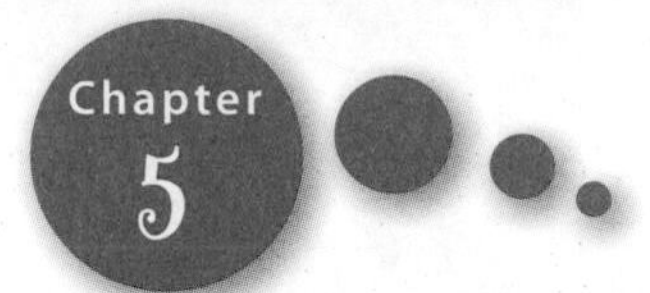

Writing Standards 3–5 (continued)

Grade 3 Students:	Grade 4 Students:	Grade 5 Students:
	Production and Distribution of Writing	
4. With guidance and support from adults, produce writing in which the development and organization are appropriate to task and purpose. (Grade-specific expectations for writing types are defined in standards 1–3 above.)	**4.** Produce clear and coherent writing in which the development and organization are appropriate to task, purpose, and audience. (Grade-specific expectations for writing types are defined in standards 1–3 above.)	**4.** Produce clear and coherent writing in which the development and organization are appropriate to task, purpose, and audience. (Grade-specific expectations for writing types are defined in standards 1–3 above.)
5. With guidance and support from peers and adults, develop and strengthen writing as needed by planning, revising, and editing. (Editing for conventions should demonstrate command of Language standards 1–3 up to and including grade 3 on page 29.)	**5.** With guidance and support from peers and adults, develop and strengthen writing as needed by planning, revising, and editing. (Editing for conventions should demonstrate command of Language standards 1–3 up to and including grade 4 on page 29.)	**5.** With guidance and support from peers and adults, develop and strengthen writing as needed by planning, revising, editing, rewriting, or trying a new approach. (Editing for conventions should demonstrate command of Language standards 1–3 up to and including grade 5 on page 29.)
6. With guidance and support from adults, use technology to produce and publish writing (using keyboarding skills) as well as to interact and collaborate with others.	**6.** With some guidance and support from adults, use technology, including the Internet, to produce and publish writing as well as to interact and collaborate with others; demonstrate sufficient command of keyboarding skills to type a minimum of one page in a single sitting.	**6.** With some guidance and support from adults, use technology, including the Internet, to produce and publish writing as well as to interact and collaborate with others; demonstrate sufficient command of keyboarding skills to type a minimum of two pages in a single sitting.
	Research to Build and Present Knowledge	
7. Conduct short research projects that build knowledge about a topic.	**7.** Conduct short research projects that build knowledge through investigation of different aspects of a topic.	**7.** Conduct short research projects that use several sources to build knowledge through investigation of different aspects of a topic.
8. Recall information from experiences or gather information from print and digital sources; take brief notes on sources and sort evidence into provided categories.	**8.** Recall relevant information from experiences or gather relevant information from print and digital sources; take notes and categorize information, and provide a list of sources.	**8.** Recall relevant information from experiences or gather relevant information from print and digital sources; summarize or paraphrase information in notes and finished work, and provide a list of sources.
9. (Begins in grade 4)	**9.** Draw evidence from literary or informational texts to support analysis, reflection, and research. **a.** Apply grade 4 Reading standards to literature (e.g., "Describe in depth a character, setting, or event in a story or drama, drawing on specific details in the text [e.g., a character's thoughts, words, or actions]."). **b.** Apply grade 4 Reading standards to informational texts (e.g., "Explain how an author uses reasons and evidence to support particular points in a text").	**9.** Draw evidence from literary or informational texts to support analysis, reflection, and research. **a.** Apply grade 5 Reading standards to literature (e.g., "Compare and contrast two or more characters, settings, or events in a story or a drama, drawing on specific details in the text [e.g., how characters interact]"). **b.** Apply grade 5 Reading standards to informational texts (e.g., "Explain how an author uses reasons and evidence to support particular points in a text, identifying which reasons and evidence support which point[s]").
	Range of Writing	
10. Write routinely over extended time frames (time for research, reflection, and revision) and shorter time frames (a single sitting or a day or two) for a range of discipline-specific tasks, purposes, and audiences.	**10.** Write routinely over extended time frames (time for research, reflection, and revision) and shorter time frames (a single sitting or a day or two) for a range of discipline-specific tasks, purposes, and audiences.	**10.** Write routinely over extended time frames (time for research, reflection, and revision) and shorter time frames (a single sitting or a day or two) for a range of discipline-specific tasks, purposes, and audiences.

Speaking and Listening Standards K–2

Kindergartners:	Grade 1 Students:	Grade 2 Students:
	Comprehension and Collaboration	
1. Participate in collaborative conversations with diverse partners about kindergarten topics and texts with peers and adults in small and larger groups. a. Follow agreed-upon rules for discussions (e.g., listening to others and taking turns speaking about the topics and texts under discussion). b. Continue a conversation through multiple exchanges.	1. Participate in collaborative conversations with diverse partners about grade 1 topics and texts with peers and adults in small and larger groups. a. Follow agreed-upon rules for discussions (e.g., listening to others with care, speaking one at a time about the topics and texts under discussion). b. Build on others' talk in conversations by responding to the comments of others through multiple exchanges. c. Ask questions to clear up any confusion about the topics and texts under discussion.	1. Participate in collaborative conversations with diverse partners about grade 2 topics and texts with peers and adults in small and larger groups. a. Follow agreed-upon rules for discussions (e.g., gaining the floor in respectful ways, listening to others with care, speaking one at a time about the topics and texts under discussion). b. Build on others' talk in conversations by linking their comments to the remarks of others. c. Ask for clarification and further explanation as needed about the topics and texts under discussion.
2. Confirm understanding of a text read aloud or information presented orally or through other media by asking and answering questions about key details and requesting clarification if something is not understood.	2. Ask and answer questions about key details in a text read aloud or information presented orally or through other media.	2. Recount or describe key ideas or details from a text read aloud or information presented orally or through other media.
3. Ask and answer questions in order to seek help, get information, or clarify something that is not understood.	3. Ask and answer questions about what a speaker says in order to gather additional information or clarify something that is not understood.	3. Ask and answer questions about what a speaker says in order to clarify comprehension, gather additional information, or deepen understanding of a topic or issue.
	Presentation of Knowledge and Ideas	
4. Describe familiar people, places, things, and events and, with prompting and support, provide additional detail.	4. Describe people, places, things, and events with relevant details, expressing ideas and feelings clearly.	4. Tell a story or recount an experience with appropriate facts and relevant, descriptive details, speaking audibly in coherent sentences.
5. Add drawings or other visual displays to descriptions as desired to provide additional detail.	5. Add drawings or other visual displays to descriptions when appropriate to clarify ideas, thoughts, and feelings.	5. Create audio recordings of stories or poems; add drawings or other visual displays to stories or recounts of experiences when appropriate to clarify ideas, thoughts, and feelings.
6. Speak audibly and express thoughts, ideas clearly.	6. Produce complete sentences when appropriate to task and situation. (See grade 1 Language standards 1 and 3 on page 26 for specific expectations.)	6. Produce complete sentences when appropriate to task and situation in order to provide requested detail or clarification. (See grade 2 Language standards 1 and 3 on page 26 for specific expectations.)

Speaking and Listening Standards 3–5

Grade 3 Student:	Grade 4 Students:	Grade 5 Students:
	Comprehension and Collaboration	
1. Engage effectively in a range of collaborative discussions (one-on-one, in groups, and teacher-led) with diverse partners on grade *3 topics and texts,* building on others' ideas and expressing their own clearly. **a.** Come to discussions prepared, having read or studied required material; explicitly draw on that preparation and other information known about the topic to explore ideas under discussion. **b.** Follow agreed-upon rules for discussions (e.g., gaining the floor in respectful ways, listening to others with care, speaking one at a time about the topics and texts under discussion). **c.** Ask questions to check understanding of information presented, stay on topic, and link their comments to the remarks of others. **d.** Explain their own ideas and understanding in light of the discussion.	**1.** Engage effectively in a range of collaborative discussions (one-on-one, in groups, and teacher-led) with diverse partners on grade 4 topics and texts, building on others' ideas and expressing their own clearly. **a.** Come to discussions prepared, having read or studied required material; explicitly draw on that preparation and other information known about the topic to explore ideas under discussion. **b.** Follow agreed-upon rules for discussions and carry out assigned roles. **c.** Pose and respond to specific questions to clarify or follow up on information, and make comments that contribute to the discussion and link to the remarks of others. **d.** Review the key ideas expressed and explain their own ideas and understanding in light of the discussion	**1.** Engage effectively in a range of collaborative discussions (one-on-one, in groups, and teacher-led) with diverse partners on grade 5 topics and texts, building on others' ideas and expressing their own clearly. **a.** Come to discussions prepared, having read or studied required material; explicitly draw on that preparation and other information known about the topic to explore ideas under discussion. **b.** Follow agreed-upon rules for discussions and carry out assigned roles. **c.** Pose and respond to specific questions by making comments that contribute to the discussion and elaborate on the remarks of others. **d.** Review the key ideas expressed and draw conclusions in light of information and knowledge gained from the discussions.
2. Determine the main ideas and supporting details of a text read aloud or information presented in diverse media and formats, including visually, quantitatively, and orally.	**2.** Paraphrase portions of a text read aloud or information presented in diverse media and formats, including visually, quantitatively, and orally.	**2.** Summarize a written text read aloud or information presented in diverse media and formats, including visually, quantitatively, and orally.
3. Ask and answer questions about information from a speaker, offering appropriate elaboration and detail.	**3.** Identify the reasons and evidence a speaker provides to support particular points.	**3.** Summarize the points a speaker makes and explain how each claim is supported by reasons and evidence.
	Presentation of Knowledge and Ideas	
4. Report on a topic or text, tell a story, or recount an experience with appropriate facts and relevant, descriptive details, speaking clearly at an understandable pace.	**4.** Report on a topic or text, tell a story, or recount an experience in an organized manner, using appropriate facts and relevant, descriptive details to support main ideas or themes; speak clearly at an understandable pace.	**4.** Report on a topic or text or present an opinion, sequencing ideas logically and using appropriate facts and relevant, descriptive details to support main ideas or themes; speak clearly at an understandable pace.
5. Add drawings or other visual displays to descriptions as desired to provide additional detail.	**5.** Add drawings or other visual displays to descriptions when appropriate to clarify ideas, thoughts, and feelings.	**5.** Include multimedia components (e.g., graphics, sound) and visual displays in presentations when appropriate to enhance the development of main ideas or themes.
5. Speak in complete sentences when appropriate to task and situation in order to provide requested detail or clarification. (See grade 3 Language standards 1 and 3 on page 28 for specific expectations.)	**5.** Differentiate between contexts that call for formal English (e.g., presenting ideas) and situations where informal discourse is appropriate (e.g., small-group discussion); use formal English when appropriate to task and situation. (See grade 4 Language standards 1 on page 28 for specific expectations.)	**5.** Adapt speech to a variety of contexts and tasks, using formal English when appropriate to task and situation. (See grade 5 Language standards 1 and 3 on page 28 for specific expectations.)
6. Speak audibly and express thoughts, ideas clearly.	**6.** Produce complete sentences when appropriate to task and situation. (See grade 1 Language standards 1 and 3 on page 26 for specific expectations.)	**6.** Produce complete sentences when appropriate to task and situation in order to provide requested detail or clarification. (See grade 2 Language standards 1 and 3 on page 26 for specific expectations.)

Language Standards K–2

Kindergartners	Grade 1 Students:	Grade 2 Students:
	Conventions of Standard English	
1. Demonstrate command of the conventions of standard English grammar and usage when writing or speaking. **a.** Print many upper- and lowercase letters. **b.** Use frequently occurring nouns and verbs. **c.** Form regular plural nouns orally by adding /s/ or /es/ (e.g., *dog, dogs; wish, wishes*). **d.** Understand and use question words (interrogatives) (e.g., *who, what, where, when, why, how*). **e.** Use the most frequently occurring prepositions (e.g., *to, from, in, out, on, off, for, of, by, with*). **f.** Produce and expand complete sentences in shared language activities.	**1.** Demonstrate command of the conventions of standard English grammar and usage when writing or speaking. **a.** Print all upper- and lowercase letters. **b.** Use common, proper, and possessive nouns. **c.** Use singular and plural nouns with matching verbs in basic sentences (e.g., He hops; We hop). **d.** Use personal, possessive, and indefinite pronouns (e.g., *I, me, my; they, them, their; anyone, everything*). **e.** Use verbs to convey a sense of past, present, and future (e.g., *Yesterday I walked home; Today I walk home; Tomorrow I will walk home*). **f.** Use frequently occurring adjectives. **g.** Use frequently occurring conjunctions (e.g., *and, but, or, so, because*). **h.** Use determiners (e.g., articles, demonstratives). **i.** Use frequently occurring prepositions (e.g., *during, beyond, toward*). **j.** Produce and expand complete simple and compound declarative, interrogative, imperative, and exclamatory sentences in response to prompts.	**1.** Demonstrate command of the conventions of standard English grammar and usage when writing or speaking. **a.** Use collective nouns (e.g., group). **b.** Form and use frequently occurring irregular plural nouns (e.g., *feet, children, teeth, mice, fish*). **c.** Use reflexive pronouns (e.g., *myself, ourselves*). **d.** Form and use the past tense of frequently occurring irregular verbs (e.g., *sat, hid, told*). **e.** Use adjectives and adverbs, and choose between them depending on what is to be modified. **f.** Produce, expand, and rearrange complete simple and compound sentences (e.g., *The boy watched the movie; The little boy watched the movie; The action movie was watched by the little boy*).
2. Demonstrate command of the conventions of standard English capitalization, punctuation, and spelling when writing. **a.** Capitalize the first word in a sentence and the pronoun *I*. **b.** Recognize and name end punctuation. **c.** Write a letter or letters for most consonant and short-vowel sounds (phonemes). **d.** Spell simple words phonetically, drawing on knowledge of sound-letter relationships.	**2.** Demonstrate command of the conventions of standard English capitalization, punctuation, and spelling when writing. **a.** Capitalize dates and names of people. **b.** Use end punctuation for sentences. **c.** Use commas in dates and to separate single words in a series. **d.** Use conventional spelling for words with common spelling patterns and for frequently occurring irregular words. **e.** Spell untaught words phonetically, drawing on phonemic awareness and spelling conventions.	**2.** Demonstrate command of the conventions of standard English capitalization, punctuation, and spelling when writing. **a.** Capitalize holidays, product names, and geographic names. **b.** Use commas in greetings and closings of letters. **c.** Use an apostrophe to form contractions and frequently occurring possessives. **d.** Generalize learned spelling patterns when writing words (e.g., cage → badge; boy → boil). **e.** Consult reference materials, including beginning dictionaries, as needed to check and correct spellings.

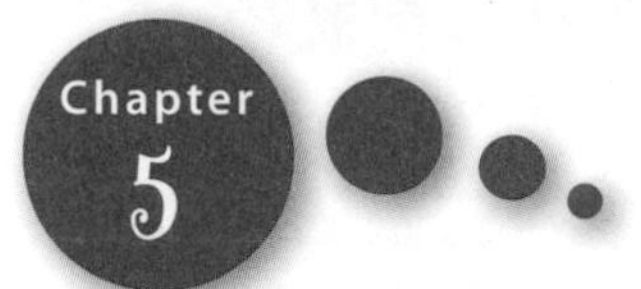

Language Standards K–2 (continued)

Kindergartners	Grade 1 Students:	Grade 2 Students:
	Knowledge of Language	
3. (Begins in grade 2)	**3.** (Begins in grade 2)	**3.** Use knowledge of language and its conventions when writing, speaking, reading, or listening. **a.** Compare formal and informal uses of English.
	Vocabulary Acquisition and Use	
4. Determine or clarify the meaning of unknown and multiple-meaning words and phrases based on *kindergarten reading and content.* **a.** Identify new meanings for familiar words and apply them accurately (e.g., knowing *duck* is a bird and learning the verb *to duck*). **b.** Use the most frequently occurring inflections and affixes (e.g., *-ed, -s, re-, un-, pre-, -ful, -less*) as a clue to the meaning of an unknown word.	**4.** Determine or clarify the meaning of unknown and multiple-meaning words and phrases based on *grade 1 reading and content,* choosing flexibly from an array of strategies. **a.** Use sentence-level context as a clue to the meaning of a word or phrase. **b.** Use frequently occurring affixes as a clue to the meaning of a word. **c.** Identify frequently occurring root words (e.g., *look*) and their inflectional forms (e.g., *looks, looked, looking*).	**4.** Determine or clarify the meaning of unknown and multiple-meaning words and phrases based on *grade 2 reading and content,* choosing flexibly from an array of strategies. **a.** Use sentence-level context as a clue to the meaning of a word or phrase. **b.** Determine the meaning of the new word formed when a known prefix is added to a known word (e.g., *happy/unhappy, tell/retell*). **c.** Use a known root word as a clue to the meaning of an unknown word with the same root (e.g., *addition, additional*). **d.** Use knowledge of the meaning of individual words to predict the meaning of compound words (e.g., *birdhouse, lighthouse, housefly; bookshelf, notebook, bookmark).* **e.** Use glossaries and beginning dictionaries, both print and digital, to determine or clarify the meaning of words and phrases.
5. With guidance and support from adults, explore word relationships and nuances in word meanings. **a.** Sort common objects into categories (e.g., shapes, foods) to gain a sense of the concepts the categories represent. **b.** Demonstrate understanding of frequently occurring verbs and adjectives by relating them to their opposites (antonyms). **c.** Identify real-life connections between words and their use (e.g., note places at school that are *colorful*). **d.** Distinguish shades of meaning among verbs describing the same general action (e.g., *walk, march, strut, prance*) by acting out the meanings.	**5.** With guidance and support from adults, demonstrate understanding of word relationships and nuances in word meanings. **a.** Sort words into categories (e.g., colors, clothing) to gain a sense of the concepts the categories represent. **b.** Define words by category and by one or more key attributes (e.g., a *duck* is a bird that swims; a *tiger* is a large cat with stripes). **c.** Identify real-life connections between words and their use (e.g., note places at home that are *cozy*). **d.** Distinguish shades of meaning among verbs differing in manner (e.g., *look, peek, glance, stare, glare, scowl*) and adjectives differing in intensity (e.g., *large, gigantic)* by defining or choosing them or by acting out the meanings.	**5.** Demonstrate understanding of word relationships and nuances in word meanings. **a.** Identify real-life connections between words and their use (e.g., describe foods that are *spicy* or *juicy*). **b.** Distinguish shades of meaning among closely related verbs (e.g., *toss, throw, hurl)* and closely related adjectives (e.g., *thin, slender, skinny, scrawny).*
6. Use words and phrases acquired through conversations, reading and being read to, and responding to texts.	**6.** Use words and phrases acquired through conversations, reading and being read to, and responding to texts, including using frequently occurring conjunctions to signal simple relationships (e.g., *because*).	**6.** Use words and phrases acquired through conversations, reading and being read to, and responding to texts, including using adjectives and adverbs to describe (e.g., *When other kids are happy that makes me happy*).

Language Standards 3–5

<table>
<tr><th>Grade 3 Students:</th><th>Grade 4 Students:</th><th>Grade 5 Students:</th></tr>
<tr><td colspan="3">Conventions of Standard English</td></tr>
<tr><td>1. Demonstrate command of the conventions of standard English grammar and usage when writing or speaking.
a. verbs, adjectives, and adverbs in general and their functions in particular sentences.
b. Form and use regular and irregular plural nouns.
c. Use abstract nouns (e.g.,childhood).
d. Form and use regular and irregular verbs.
e. Form and use the simple (e.g., I walked; I walk; I will walk) verb tenses.
f. Ensure subject-verb and pronoun-antecedent agreement.
g. Form and use comparative and superlative adjectives and adverbs, and choose between them depending on what is to be modified.
h. Use coordinating and subordinating conjunctions.
i. Produce simple, compound, and complex sentences.</td><td>1. Demonstrate command of the conventions of standard English grammar and usage when writing or speaking.
a. Use relative pronouns (who, whose, whom, which, that) and relative adverbs (where, when, why).
b. Form and use the progressive (e.g., I was walking; I am walking; I will be walking) verb tenses.
c. Use modal auxiliaries (e.g., can, may, must) to convey various conditions.
d. Order adjectives within sentences according to conventional patterns (e.g., a small red bag rather than a red small bag).
e. Form and use prepositional phrases.
f. Produce complete sentences, recognizing and correcting inappropriate fragments and run-ons.
g. Correctly use frequently confused words (e.g., to, too, two; there, their).</td><td>1. Demonstrate command of the conventions of standard English grammar and usage when writing or speaking.
a. Explain the function of conjunctions, prepositions, and interjections in general and their function in particular sentences.
b. Form and use the perfect (e.g., I had walked; I have walked; I will have walked) verb tenses.
c. Use verb tense to convey various times, sequences, states, and conditions.
d. Recognize and correct inappropriate shifts in verb tense.
e. Use correlative conjunctions (e.g., either/or, neither/nor).</td></tr>
<tr><td>2. Demonstrate command of the conventions of standard English capitalization, punctuation, and spelling when writing.
a. Capitalize appropriate words in titles.
b. Use commas in addresses.
c. Use commas and quotation marks in dialogue.
d. Form and use possessives.
e. Use conventional spelling for high-frequency and other studied words and for adding suffixes to base words (e.g., sitting, smiled, cries, happiness).
f. Use spelling patterns and generalizations (e.g., word families, position-based spellings, syllable patterns, ending rules, meaningful word parts) in writing words.
g. Consult reference materials, including beginning dictionaries, as needed to check and correct spellings.</td><td>2. Demonstrate command of the conventions of standard English capitalization, punctuation, and spelling when writing.
a. Use correct capitalization.
b. Use commas and quotation marks to mark direct speech and quotations from a text.
c. Use a comma before a coordinating conjunction in a compound sentence.
d. Spell grade-appropriate words correctly, consulting references as needed.</td><td>2. Demonstrate command of the conventions of standard English capitalization, punctuation, and spelling when writing.
a. Use punctuation to separate items in a series.
b. Use a comma to separate an introductory element from the rest of the sentence.
c. Use a comma to set off the words yes and no (e.g., Yes, thank you), to set off a tag question from the rest of the sentence (e.g., It's true, isn't it?), and to indicate direct address (e.g., Is that you, Steve?).
d. Use underlining, quotation marks, or italics to indicate titles of works.
e. Spell grade-appropriate words correctly, consulting references as needed.</td></tr>
</table>

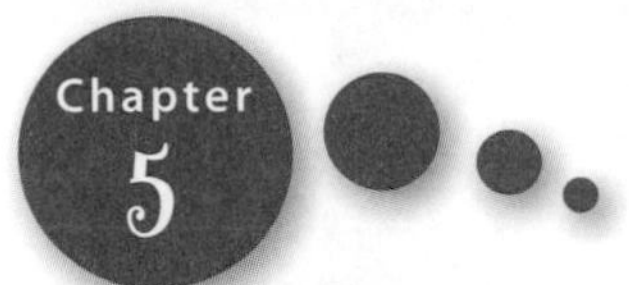

Language Standards 3–5 (continued)

Grade 3 Students:	Grade 4 Students:	Grade 5 Students:
	Knowledge of Language	
3. Use knowledge of language and its conventions when writing, speaking, reading, or listening. **a.** Choose words and phrases for effect.* **b.** Recognize and observe differences between the conventions of spoken and written standard English.	**3.** Use knowledge of language and its conventions when writing, speaking, reading, or listening. **a.** Choose words and phrases to convey ideas precisely.* **b.** Choose punctuation for effect.* **c.** Differentiate between contexts that call for formal English (e.g., presenting ideas) and situations where informal discourse is appropriate (e.g., small-group discussion).	**3.** Use knowledge of language and its conventions when writing, speaking, reading, or listening. **a.** Expand, combine, and reduce sentences for meaning, reader/listener interest, and style. **b.** Compare and contrast the varieties of English (e.g., dialects, registers) used in stories, dramas, or poems.
	Vocabulary Acquisition and Use	
4. Determine or clarify the meaning of unknown and multiple-meaning word and phrases based on grade *3 reading and content,* choosing flexibly from a range of strategies. **a.** Use sentence-level context as a clue to the meaning of a word or phrase. **b.** Determine the meaning of the new word formed when a known affix is added to a known word (e.g., *agreeable/disagreeable, comfortable/uncomfortable, care/careless, heat/preheat*). **c.** Use a known root word as a clue to the meaning of an unknown word with the same root (e.g., *company, companion*). **d.** Use glossaries or beginning dictionaries, both print and digital, to determine or clarify the precise meaning of key words and phrases.	**4.** Determine or clarify the meaning of unknown and multiple-meaning words and phrases based on *grade 4 reading and content,* choosing flexibly from a range of strategies. **a.** Use context (e.g., definitions, examples, or restatements in text) as a clue to the meaning of a word or phrase. **b.** Use common, grade-appropriate Greek and Latin affixes and roots as clues to the meaning of a word (e.g., *telegraph, photograph, autograph*). **c.** Consult reference materials (e.g., dictionaries, glossaries, thesauruses), both print and digital, to find the pronunciation and determine or clarify the precise meaning of key words and phrases.	**4.** Determine or clarify the meaning of unknown and multiple-meaning words and phrases based on *grade 5 reading and content,* choosing flexibly from a range of strategies. **a.** Use context (e.g., cause/effect relationships and comparisons in text) as a clue to the meaning of a word or phrase. **b.** Use common, grade-appropriate Greek and Latin affixes and roots as clues to the meaning of a word (e.g., *photograph, photosynthesis*). **c.** Consult reference materials (e.g., dictionaries, glossaries, thesauruses), both print and digital, to find the pronunciation and determine or clarify the precise meaning of key words and phrases.
5. Demonstrate understanding of word relationships and nuances in word meanings. **a.** Distinguish the literal and nonliteral meanings of words and phrases in context (e.g., *take steps*). **b.** Identify real-life connections between words and their use (e.g., describe people who are *friendly* or *helpful*). **c.** Distinguish shades of meaning among related words that describe states of mind or degrees of certainty (e.g., *knew, believed, suspected, heard, wondered*).	**5.** Demonstrate understanding of figurative language, word relationships, and nuances in word meanings. **a.** Explain the meaning of simple similes and metaphors (e.g., *as pretty as a picture*) in context. **b.** Recognize and explain the meaning of common idioms, adages, and proverbs. **c.** Demonstrate understanding of words by relating them to their opposites (antonyms) and to words with similar but not identical meanings (synonyms).	**5.** Demonstrate understanding of figurative language, word relationships, and nuances in word meanings. **a.** Interpret figurative language, including similes and metaphors, in context. **b.** Recognize and explain the meaning of common idioms, adages, and proverbs. **c.** Use the relationship between particular words (e.g., synonyms, antonyms, homographs) to better understand each of the words.
6. Acquire and use accurately grade-appropriate conversational, general academic, and domain-specific words and phrases, including those that signal spatial and temporal relationships (e.g., *After dinner that night we went looking for them*).	**6.** Acquire and use accurately grade-appropriate general academic and domain-specific words and phrases, including those that signal precise actions, emotions, or states of being (e.g., *quizzed, whined, stammered*) and that are basic to a particular topic (e.g., *wildlife, conservation,* and *endangered* when discussing animal preservation).	**6.** Acquire and use accurately grade-appropriate general academic and domain-specific words and phrases, including those that signal contrast, addition, and other logical relationships (e.g., *however, although, nevertheless, similarly, moreover, in addition*).

Freezing and Melting

Description

Frozen treats provide a familiar and fun context for learning about changes in matter. Through engaging read-alouds and some cool activities (pun intended) with Popsicles and ice cream, students learn about solids, liquids, freezing, and melting.

Suggested Grade Levels: K–2

LESSON OBJECTIVES *Connecting to the Framework*

PHYSICAL SCIENCES

Core Idea PS1: Matter and Its Interactions

PS1.A: Structure and Properties of Matter

By the end of grade 2: Different kinds of matter exist (e.g., wood, metal, water), and many of them can be either solid or liquid, depending on temperature.

PS1.B: Chemical Reactions

By the end of grade 2: Heating or cooling a substance may cause changes that can be observed. Sometimes these changes are reversible (e.g., melting and freezing), and sometimes they are not (e.g., baking a cake, burning fuel).

Featured Picture Books

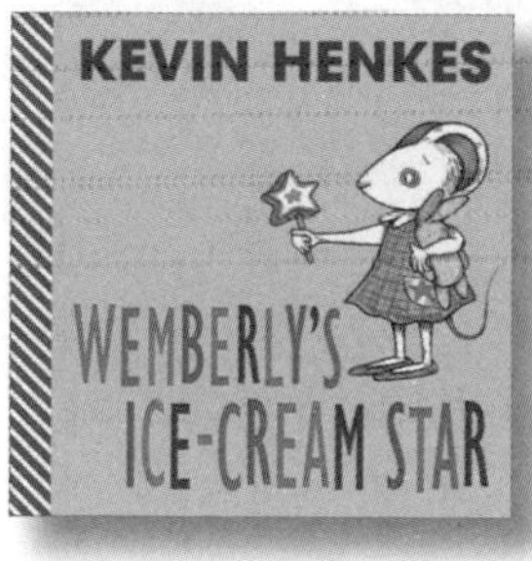

TITLE: ***Wemberly's Ice-Cream Star***
AUTHOR: **Kevin Henkes**
ILLUSTRATOR: **Kevin Henkes**
PUBLISHER: **Greenwillow Books**
YEAR: **2003**
GENRE: **Story**
SUMMARY: *Wemberly wants to share her ice-cream star with Petal, so she waits until it melts and then makes ice-cream star soup for them to share.*

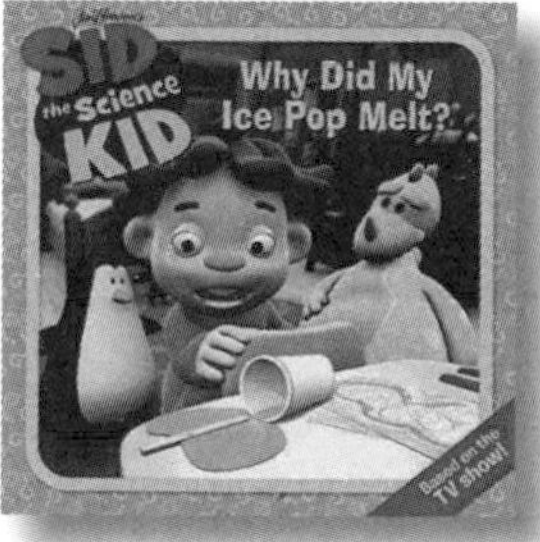

TITLE: ***Why Did My Ice Pop Melt?***
AUTHOR: **Susan Korman**
PUBLISHER: **HarperFestival**
YEAR: **2010**
GENRE: **Narrative Information**
SUMMARY: *Based on the PBS television series* Sid the Science Kid, *this book explains why things melt and introduces the concept of a reversible change.*

Time Needed

This lesson will take several class periods. Suggested scheduling is as follows:

Day 1: **Engage** with *Wemberly's Ice Cream Star* Read-aloud, **Explore** with Popsicle Soup: Part 1, and **Explain** with *Why Did My Ice Pop Melt?* Read-Aloud

Day 2: Liquid and Solid Demonstrations and Popsicle Soup: Part 2

Day 3: **Elaborate** with Ice Cream Ingredient Exploration and Making Ice Cream

Day 4: **Evaluate** with The Day We Made Ice Cream

Materials

For making Popsicle soup (per student)

- Store-bought or homemade freezer pop
- Bowl
- Spoon

For Liquid and Solid Demonstrations

- Food coloring
- Water
- Clear plastic bottle
- Clear container with a different shape from the bottle
- Pencil

For Making Ice Cream (per group of four students)

- 2 gallon-size zippered plastic bags (heavy duty)
- 1 quart-size zippered plastic bag (heavy duty)
- 4 cups ice
- ¼ cup salt
- 1 cup whole milk
- 1 teaspoon vanilla extract
- 2 tablespoons sugar
- 4 small cups
- 4 spoons
- 4 pairs of winter gloves (*Note:* If you ask students to bring these from home, be sure to have a few extras for anyone who forgets to bring them.)
- Hand lens
- Alternative treats for students with dairy allergies

SAFETY

- Check with the school nurse regarding student medical issues (e.g., allergies) and how to deal with them.
- Have students wash their hands with soap and water upon completing the activity (before and after when consuming food).
- When making food to be eaten (e.g., ice cream), make sure that all surfaces and equipment for making the food have been sanitized.
- When working with cool or cold liquids/solids, have students use appropriate personal protective equipment (PPE), including thermal gloves, eye protection, and aprons.

Student Pages

- Popsicle Soup
- The Day We Made Ice Cream (copy p. 62 back-to-back with p. 63)

Background

According to *A Framework for K–12 Science Education,* students should build some foundational ideas about matter and its interactions in the early years of school. Specifically, they should understand that different kinds of matter exist and that they can be different forms based on temperature. They should also have opportunities to observe that heating and cooling matter can cause changes; some of these changes are reversible (e.g., freezing and melting) and some are not (baking a cake, burning fuel). This lesson uses the familiar context of frozen treats to give students experience with solids, liquids, freezing, and melting and opportunities to use that vocabulary.

Making ice cream in the classroom is a fun way to explore freezing and melting. The main ingredients are simple: whole milk, sugar, and vanilla. The key is getting the ingredient mixture cold enough to become solid ice cream. In the lesson, we suggest you place all the ingredients in a quart-size zippered bag, and then place that bag into a gallon-size zippered bag containing ice and either rock salt or kosher salt. The salt lowers the freezing point of water from its usual freezing point of 0°C (32°F) to −2°C (28°F), making the ice-salt-water mixture in the outside bag much colder than if it were just ice alone. This very cold outer mixture causes the liquid milk mixture to freeze and become solid. Shaking the bag distributes the cold outer mixture so that it makes better contact with the inner bag.

Wemberly's Ice-Cream Star Read-Aloud

Inferring

> Connecting to the Common Core
> **Reading: Literature**
> Key Ideas and Details: K.1, 1.1, 2.1

Show students the cover of *Wemberly's Ice-Cream Star* and introduce the author/illustrator. *Ask*

? What do you think this story is going to be about? (Answers will vary.)

? Have you ever read any other books about Wemberly or any other books by Kevin Henkes? (Students may have read popular books by Kevin Henkes, such as *Wemberly Worried, Owen, Chrysanthemum,* or *Julius, the Baby of the World.*)

Questioning

Read the book aloud. *Ask*

? What happened to Wemberly's ice-cream star? (It melted.)

? Why did it melt? (It got warmer.)

? Why did Wemberly want her ice-cream star to melt? (She wanted to share it with Petal.)

? What is your favorite frozen treat?

? How many of you like Popsicles?

? Has anyone ever had Popsicle soup?

explore

Popsicle Soup: Part 1

Give each student the Popsicle Soup student page. Tell them that after they write their names at the top, they will receive a bowl and a Popsicle. *Ask*

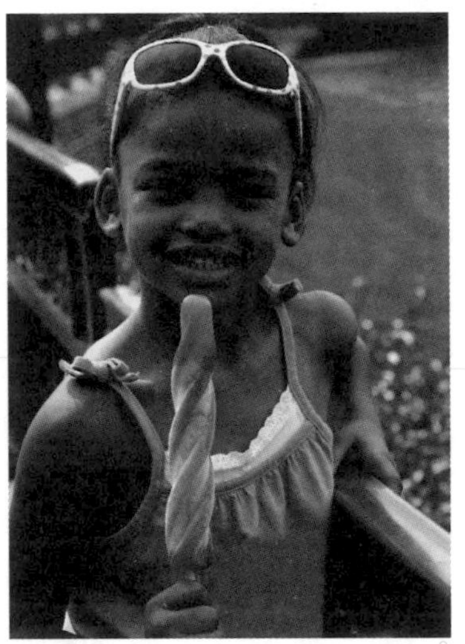

POPSICLE "BEFORE"

? How can we make Popsicle soup? (Wait for the Popsicle to melt.)

On part 1 of the Popsicle Soup student page, have students draw a picture of the Popsicle in the "Before" section of the student page. Tell students that when they are finished with their drawing, they can take a small bite of their Popsicle and then come to the reading carpet or corner. *Ask*

? How did your Popsicle taste?

? What flavor was it?

? How did it feel on your tongue? (cold, hard)

Explain to students that you are going to read a story while they wait to eat their Popsicle soup.

explain

Why Did My Ice Pop Melt? Read-Aloud

Making Connections

Connecting to the Common Core
Reading: Informational Text
KEY IDEAS AND DETAILS: K.1, 1.1, 2.1

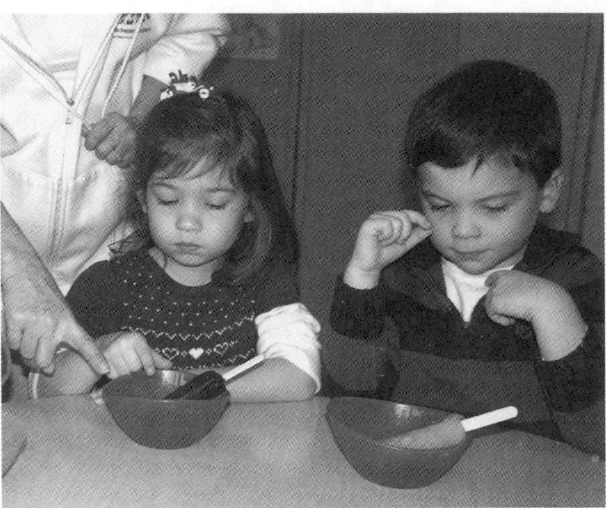

MAKING POPSICLE SOUP

Show students the cover of *Why Did My Ice Pop Melt?* and explain that "ice pop" is another name for a Popsicle. *Ask*

? Have you ever read a *Sid the Science Kid* book or seen the show on television?

Explain that the character Sid is a kid who is very curious about science.

Read the story aloud, stopping periodically to ask guiding questions, such as:

Page 4: What does "melted" mean? (turned into a liquid)

Page 5: Why do frozen things melt? (They get warmer.)

Page 9: Why aren't the ice pops ready yet? (They need more time to get colder.)

Page 14: Do you think it is possible to freeze liquid back into ice?

Page 15: What does the word *reverse* mean? (to go back) What is it called when something can change back and forth? (reversible change)

Page 17: What do you think Sid and his friends can do to make the ice melt faster? (Answers will vary but may involve making the ice warmer or breaking it in to smaller pieces.)

Page 21: Why did the water make the block of ice melt faster? (It was warm.)

Tell students to go back to their desks to check on their Popsicle soup. They should draw it in the "After" section of the student page. Tell them to raise their hand when they finish their drawing, and you will bring them a spoon so they can eat their Popsicle soup. Tell them they will be doing part 2 of the student page later.

Liquid and Solid Demonstrations

> Connecting to the Common Core
> **Language**
> Vocabulary Acquisition and Use: K.6, 1.6, 2.6

Reread page 14 of *Why Did My Ice Pop Melt?* where the teacher uses the word "liquid." Then create a T-chart on the board with the word "liquid" on one side. *Ask*

- ? What is a liquid?

  Do the following demonstration of a liquid:

  Hold up a plastic bottle filled with colored water. Then hold up an empty clear container with a different shape. Ask students to predict what shape the colored water will take when you pour it into the other container. Demonstrate, and repeat a few times. Then explain that a liquid is a kind of matter that can pour, and that it takes the shape of whatever container it is in.

Ask

- ? What are some other examples of liquids? (milk, juice, and so on)

List correct student examples on the T-chart.

Ask

- ? Is a frozen Popsicle a liquid? (no)
- ? If it's not a liquid, what is it? (a solid)

Write the word "solid" on the other side of the T-chart. Then do the following demonstration of a solid:

Hold up a pencil and one of the empty clear containers. Ask students to predict what shape the pencil will take when you put it into the container. Demonstrate by placing the pencil into the container. Then explain that solids have their own shape and do not take the shape of the container they are in.

Ask

- ? What are some examples of solids? (ice, pencil, book, and so on)

  List correct student examples on the T-chart.

Ask

- ? What is it called when something changes from a solid to a liquid? (melting)
- ? Can you think of other things that melt? (candles, chocolate, cheese, ice cream, and so on)
- ? How do you make things melt? (Add heat, make them warmer.)

  Explain that some things do not melt when they are heated, but instead will burn if they get hot enough or set on fire. *Ask*

- ? Can you think of something else that doesn't melt when it is heated, but burns instead? (wood, paper, etc.)

Ask

- ? Would burning be a "reversible change"? (no)
- ? Why not? (It can't go back to what it was before.)

Popsicle Soup: Part 2

Tell the students to think back to when they ate their Popsicle soup. *Ask*

- ? Did the soup have the same flavor as the Popsicle? (yes)
- ? Did it feel the same in your mouth? (no)
- ? What was the difference? (It was liquid, it was runny, it was not as cold)

Observing ice cream ingredients

Explain that the Popsicle soup was still the same "stuff" they started with. It was just in a liquid form. *Ask*

- ? Would it be possible to turn the Popsicle soup back into a Popsicle? (yes)
- ? What would we have to do? (Freeze it.)
- ? What is that kind of change called, where it can change back and forth? (reversible change)

Next, have students complete part 2 of the Popsicle Soup student page by filling in the blanks with the correct words, so that the sentences read:

My Popsicle started out as a solid.

My Popsicle got warmer and turned into a liquid.

Changing from a solid to a liquid is called melting.

Encourage the students to take their papers home and explain to someone how they made Popsicle soup and read their completed sentences aloud to that person.

elaborate

In advance, ask each student to bring in a pair of winter gloves.

Ice Cream Ingredient Exploration

Ask

- ? How is ice cream made?
- ? Do you think we could make ice cream right here in our classroom?
- ? What materials do you think we would need to make ice cream right here, right now?

Have students explain their reasoning for each ingredient they suggest.

Tell students that you have the ingredients to make vanilla ice cream and you would like them to use their senses of sight and smell to figure out what they are. Give each group of four students small cups containing small amounts of sugar, vanilla, whole milk, and ice. Allow time for them to observe the ingredients, without tasting, and discuss what they might be. Then identify each substance and discuss its purpose. For example, sugar makes the ice cream sweet, vanilla adds flavor, ice makes it freeze.

Have the students sort the ingredients into two groups: liquid (milk and vanilla) and solid (ice and sugar). Students might have trouble determining if sugar is a solid or liquid because it pours,

Making ice cream

so explain that some solids are so tiny that they can be poured. Have students observe one grain of sugar with a hand lens to see that it is a solid cube. It has its own shape, therefore it is a solid.

Making Ice Cream

? What do we need to do to turn this mixture into ice cream? (Make it cold.)

> **NOTE:** The instructions in this activity are for making ice cream in groups of four. You may want to have parent volunteers help out or invite some older student "buddies" to help. If you are not able to get assistance, you can adapt this activity to be a demonstration.

- Measure the milk, vanilla, and sugar and place them into a small zippered bag for each group of four students. Squeeze out as much air as possible and **seal the bag carefully and completely**. Too much air left inside may force the bag open during shaking.
- Fill a gallon-size zippered bag about half full of ice and add ¼ cup of salt. Put this bag inside the other gallon-size zippered bag. (Double bagging is recommended to prevent leaks.) Explain to students that the salt is added to make the ice even colder.
- Place the small zippered bag down into a large zippered bag. Make sure the small bag is pushed down into the ice and salt. **Seal the large bag tightly**.
- The bag gets very cold, so have the students put on winter gloves.
- Give each team a bag and have them pass, shake, and flip the bag for about 10 minutes. Tell students not to open the large bag to check the ice cream because it may not seal properly afterward.
- Take the small bag out of the large bag of ice and wipe it off with paper towels to get the salt water off of the outside of the bag.
- Open the bag and spoon the ice cream into small cups. (*Note:* To avoid getting salty water into the ice cream, it helps to fold the top of the small bag outward before scooping out the ice cream.)

Ask

? Should we eat our ice cream now? (Yes!)

? What would happen if we waited until later to eat it? (It would melt.)

Provide an alternative treat for students with dairy allergies, then allow everyone to enjoy their treats!

evaluate

The Day We Made Ice Cream

Writing

> **Connecting to the Common Core**
> **Writing**
> Research to Build and Present Knowledge: K.7, 1.7, 2.7
> **Language**
> Vocabulary Acquisition and Use: K.6, 1.6, 2.6

Give each student a copy of The Day We Made Ice Cream student page. Have each student write the correct word to complete each sentence and then illustrate each sentence in the box. Encourage students to go back to each page and add more details to the story to make it more personal. For example, a student might add a description of how the ice cream tasted on the last page.

The sentences should read:

First we mixed milk, sugar, and vanilla in a bag. It was a liquid.

Next we added solid ice.

Then we shook the bag to make the liquid freeze into a solid.

Last we scooped out the solid ice cream.

We had to eat our ice cream before it melted.

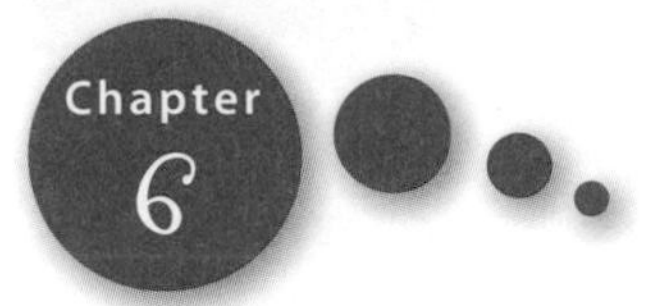

Inquiry Place

Have students brainstorm questions about solids and liquids. Examples of such questions include

- ? Does the amount of a liquid affect how quickly it will freeze into a solid? Test it!
- ? Does the amount of a solid affect how quickly it will melt into a liquid? Test it!
- ? Does temperature affect how quickly a solid will melt into a liquid? Test it!
- ? How cold does the temperature have to be for water to freeze into a solid? Test it!

Then have students select a question to investigate as a class, or have groups of students vote on the question they want to investigate as a team. After they make predictions, have them conduct an experiment to test their predictions. Students can present their findings at a poster session or gallery walk.

More Books to Read

Gibbons, G. 2006. *Ice cream: The full scoop*. New York: Holiday House.

Summary: This informative book explains the history of ice cream and how it is made.

Hansen, A. 2012. *Matter comes in all shapes*. Vero Beach, FL: Rourke.

Summary: Simple text and photographs explain the differences between solids, liquids, and gases.

Hansen, A. 2012. *Solid or liquid?* Vero Beach, FL: Rourke.

Summary: Simple text and photographs explain the differences between solids and liquids.

Royston, A. 2008. *Solids, liquids, and gases*. Chicago, IL: Heinemann.

Summary: From the *My World of Science* series, this book introduces the three states of matter and their properties. It also includes information on freezing, melting, and dissolving.

Schuh, M. 2012. *All about matter*. Mankato, MN: Capstone Press.

Summary: Simple text and photographs provide a brief introduction to matter and its properties.

Willems, M. 2010. *Should I share my ice cream?* New York: Hyperion Books for Children.

Summary: From the *Elephant and Piggie* series, this book follows Gerald the elephant as he makes a big decision: Should he share his ice cream? He waits too long and it melts, but Piggie brings more and saves the day.

Zoehfeld, K. W. 1998. *What is the world made of? All about solids, liquids, and gases*. New York: HarperCollins.

Summary: From the *Let's-Read-and-Find-Out Science* series, this book gives examples of each state of matter and some simple activities that demonstrate the attributes of each.

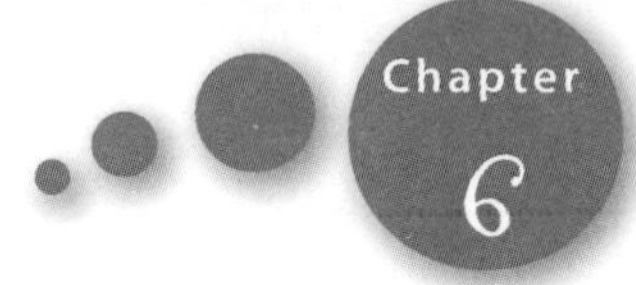

Name: ______________________________

Popsicle Soup

Part 1

Draw a picture of your Popsicle before and after it became Popsicle soup in the boxes below.

Before **After**

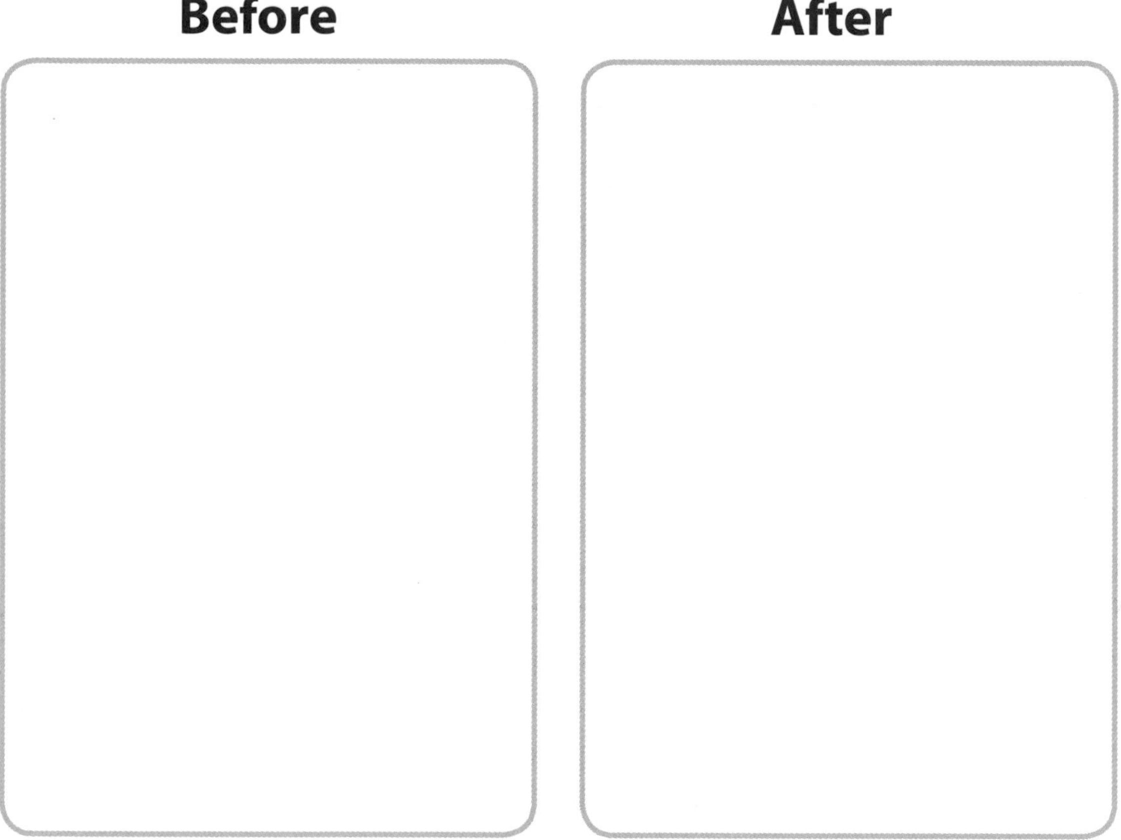

Part 2

My Popsicle started out as a ______________________________.
(solid or liquid)

My Popsicle got warmer and turned into a ____________________.
(solid or liquid)

Changing from a solid to a liquid is called ____________________.
(freezing or melting)

We had to eat our ice cream before it

____________________________.

(froze or melted)

The Day We Made Ice Cream

By ____________________________

First, we mixed milk, sugar, and vanilla in a bag. It was a ________________.
(solid or liquid)

Next, we added ________________ ice.
(solid or liquid)

Then, we shook the bag of ice to make the liquid ________________ into a solid.
(freeze or melt)

Last, we scooped out the ________________ ice cream.
(solid or liquid)

Float Your Boat

Description

How does a cruise ship, made of tons of steel, stay afloat on the water? In this lesson, students make boats out of aluminum foil, learn how gravity and buoyancy affect boats, and apply what they have learned about these forces to come up with a foil boat design that will hold the most pennies.

Suggested Grade Levels: 3–5

LESSON OBJECTIVES *Connecting to the Framework*

PHYSICAL SCIENCES

Core Idea PS2: Motion and Stability: Forces and Interactions

PS2.A: Forces and Motion

By the end of grade 5: Each force acts on one particular object and has both a strength and a direction. An object at rest typically has multiple forces acting on it, but they add to give zero net force on the object. Forces that do not sum to zero can cause changes in the object's speed or direction of motion.

PS2.B: Types of Interactions

By the end of grade 5: Objects in contact exert forces on each other (friction, elastic pushes and pulls). Electric, magnetic, and gravitational forces between a pair of objects do not require that the objects be in contact—for example, magnets push or pull at a distance. The sizes of the forces in each situation depend on the properties of the objects and their distances apart and, for forces between two magnets, on their orientation relative to each other. The gravitational force of Earth acting on an object near Earth's surface pulls that object toward the planet's center.

ENGINEERING, TECHNOLOGY, AND APPLICATIONS OF SCIENCE

Core Idea ETS1: Engineering Design

ETS1.C: Optimizing the Design Solution

By the end of grade 5: Different solutions need to be tested in order to determine which of them best solves the problem, given the criteria and the constraints.

Featured Picture Books 

TITLE: *Toy Boat*
AUTHOR: **Randall de Sève**
ILLUSTRATOR: **Loren Long**
PUBLISHER: **Philomel Books**
YEAR: **2007**
GENRE: **Story**
SUMMARY: *A toy boat gets separated from its owner and has an adventure on the high seas.*

TITLE: *Captain Kidd's Crew Experiments With Sinking and Floating*
AUTHOR: **Mark Weakland**
ILLUSTRATOR: **Troy Cummings**
PUBLISHER: **Picture Window Books**
YEAR: **2012**
GENRE: **Narrative Information**
SUMMARY: *Tells the story of a pirate and his crew experimenting with sinking and floating. Captain Kidd explains the forces of gravity and buoyancy and how they affect boats.*

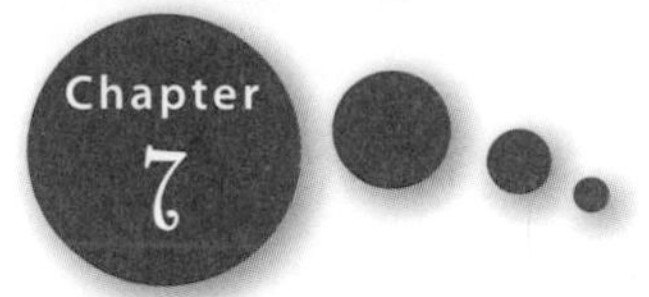

Time Needed

This lesson will take several class periods. Suggested scheduling is as follows:

Day 1: **Engage** with *Toy Boat* Read-Aloud, **Explore** with Float Your Boat, and **Explain** with Float Your Boat Discussion

Day 2: **Explain** with *Captain Kidd's Crew Experiments With Sinking and Floating* Read-Aloud and **Elaborate** with A Ship for Captain Kidd

Day 3: **Evaluate** with Letter to Captain Kidd

Materials

For Float Your Boat (per group of three to five students)

- 15 cm × 15 cm piece of heavy-duty aluminum foil (1 per student)
- Tub of water
- 50 pennies
- Paper towels

For Stop and Try It (per class)

- Balloon
- Tub of water
- 15 cm × 15 cm square of heavy-duty aluminum foil
- (Optional) Balance to compare the weight of the pieces of foil

For A Ship for Captain Kidd (per group of three to five students)

- Roll of heavy-duty aluminum foil
- Tub of water
- Metric ruler
- Roll of masking tape
- Scissors
- Blank 3"× 5" index card
- Markers
- Bendable straw
- 50 pennies

SAFETY

- Be careful to quickly wipe up any spilled water, oil, or other liquid on the floor. This is a slip/fall hazard, which can result in a serious injury.
- When working with glassware, metersticks, wires, projectiles, tools, straws, or other solid hazards, have students use appropriate personal protective equipment (PPE), including safety glasses or goggles that meet the ANSI Z87.1 standard.
- Use caution in working with sharp items like scissors, wires, open paper clips, screwdrivers, metal pans and soda cans, wood, and glass (including thermometers). They can cut or puncture skin.
- When working with plastic bags or balloons, remind students to keep them away from their mouths. These are potential breathing and/or choking hazards.

Student Pages

- Float Your Boat
- A Ship for Captain Kidd
- Letter to Captain Kidd

Background

A Framework for K–12 Science Education suggests that students should be involved in engineering design challenges, which require the application of scientific principles. In this lesson, students are given the opportunity to do just that. In order to design a solution to an engineering challenge, students must learn some basic scientific principles that affect sinking and floating. Students learn about the forces of gravity and buoyancy as well as the relationship between buoyant force and water displacement. They then apply what they have learned by designing a foil boat that not only will float but will hold the greatest number of pennies before it sinks.

Determining whether an object will sink or float depends on its density, or its weight compared with its size. If an object is more dense than the liquid or gas that it is in, it will sink. If it is less dense, it will float. This is why a cork floats and a penny sinks. A cork is less dense than water, so it floats. A penny is denser than water, so it sinks. But how does a ship made of tons of steel float on the water? We know that the force of gravity is pulling down on that steel, so why doesn't it end up at the bottom of the ocean? It's because of a force called buoyancy that is pushing up on the ship. Buoyancy is the force of liquid pushing upward. A steel boat is shaped so as to increase the amount of water it displaces. If it displaces enough water, then the buoyant force is large enough to keep it afloat. Another way to look at it is to include all the air contained in the boat as part of the boat. With enough air, the overall density of the boat is actually less than the density of water. In this lesson when students make boats with aluminum foil, they will see that a tight ball of aluminum foil sinks, but the same amount of foil will float if it is spread out so that it comes in contact with more water. Because the foil boat displaces more water than the small foil ball, there is more buoyant force pushing up on the boat, thus keeping it afloat.

So, there are two ways to think about floating and sinking. One way is to compare densities. This works well when determining whether certain kinds of materials will float or sink or if certain liquids will float on top of one another. However, this lesson focuses on a second way to think about sinking and floating—the forces of buoyancy and gravity. Even though you can explain why a ship floats using the concept of density (the density of the ship is less than the water if you include the air inside the ship in the equation), this lesson focuses on the concept of the size and direction of the particular forces involved in floating and sinking: buoyancy and gravity, rather than density. *A Framework for K–12 Science Education* suggests that by the end of grade 5 students should understand that objects in contact exert forces on one another but that some forces do not need to be in contact to act on an object (e.g., gravity and magnetism.) They should also learn that each force that acts on an object has a strength (magnitude) and a direction, and the motion of an object depends on the strength (magnitude) and direction of the forces acting on it.

Toy Boat Read-Aloud

Show students the cover of *Toy Boat* and tell them that you are going to read the story to them twice. The first time, the purpose of the read-aloud is to hear the story and enjoy it. Read the book aloud, then *ask*

? Is this book fiction or nonfiction? (fiction)

? How do you know? (The toy boat thinks and has feelings and real boats do not.)

> Connecting to the Common Core
> **Reading: Literature**
> Key Ideas and Details: 3.1, 4.1, 5.1

Tell students that you are going to read the story again, and this time you would like them to compare the different types of boats in the story—what's different about them and what's the same. Explain that the watertight body of a ship or boat is called the *hull*, the front of the boat is called the *bow*, and the back is called the *stern*. Encourage students to use this vocabulary when describing the boat shapes.

Then read the book aloud, pausing to point out the various pictures of boats; their names; sizes; and hull, bow, and stern shapes. *Ask*

- ? How are the different boats shaped? (Answers will vary, but students should note that they are all hollow on the inside. Some boats come to a point at the bow, some are squared off at the stern, and so on.)

After reading, tell students that the author of this book, Randall de Sève, was inspired to write this story by a boat that she and her daughter made from a can, a cork, a toothpick, and some white tape. Likewise, the illustrator, Loren Long, has fond memories of creating paper boats with his brothers on rainy days and following them as they floated on the puddles.

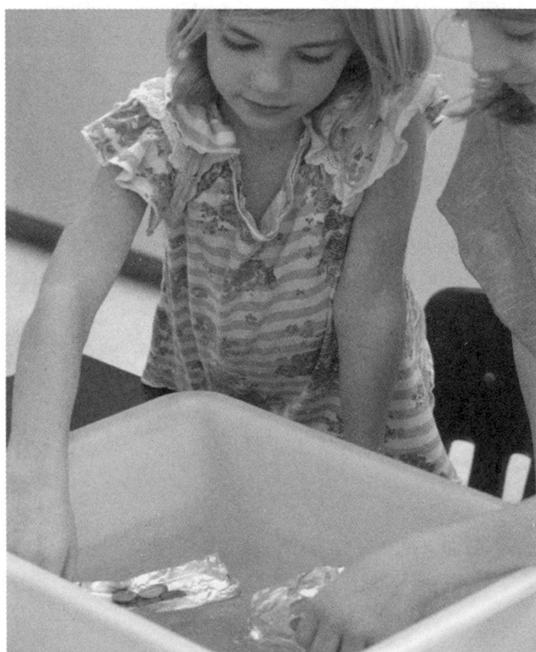

Testing the Boats

Ask

- ? Have you ever made a boat before?
- ? What kinds of materials would be good for making a toy boat? (materials that are waterproof, not too heavy, and so on)
- ? What are most real boats made of? (steel, wood, or fiberglass)
- ? What are boats designed to do? (carry people or things across bodies of water)
- ? Why do boats float? (Answers will vary according to students' preconceptions.)

explore

Float Your Boat

Give each student a Float Your Boat student page and a 15 cm × 15 cm piece of heavy-duty aluminum foil, and tell them that they will be making their own boats out of the aluminum foil. Students will be working in groups of three to five, but each student will build his or her own boat. When they are done constructing their boats, they can begin testing to see how many pennies each boat will hold by gently placing the pennies in the boat one by one until it sinks. Tell students not to count the last penny that made the boat sink. Have them record the number of pennies each boat held on the student page. After every boat is tested, they should record the name of the person whose boat held the most pennies in their group, draw a picture of that boat, and explain why they think that boat held the most.

explain

Float Your Boat Discussion

After all groups have finished, bring the students together with their papers so they can explain what happened in their groups. *Ask*

? What shape was the boat that held the most pennies in your group? (Generally boats with a flat hull and high sides work best.)

? How was that boat different from the others?

? Did any other boats in your group come close to the amount of pennies it held? Why is it important to place the pennies in the boat gently? (Dropping the pennies adds an extra force exerted on the boat. That extra force might sink the boat.)

? Does it matter where you place the pennies in the boat? (Students should have figured out that if they spread out the pennies, the boat will hold more.)

Captain Kidd's Crew Experiments With Sinking and Floating Read-Aloud

Connecting to the Common Core
Reading: Informational Text
KEY IDEAS AND DETAILS: 3.1, 4.1, 5.1

Tell students that they will have an opportunity to revise their foil boat design and give it another try with the pennies. But, first, you have another book to share with them.

Questioning

Show them the cover of *Captain Kidd's Crew Experiments With Sinking and Floating.*

Ask

? What things are floating in this picture? (the ship, the bottle, the rowboat)

? What things are sinking? (the treasure)

? Why do you think some things sink and some things float? (Answers will vary according to students' preconceptions.)

Tell them that this book might help them understand more about floating and sinking.

STOP and TRY IT

While reading aloud *Captain Kidd's Crew Experiments With Sinking and Floating,* pause where noted to ask questions about the reading and try activities that demonstrate the scientific principles involved in sinking and floating.

First, read aloud pages 1–4.

STOP

After reading page 4, *ask*

? What force pulls objects down toward the center of the earth? (gravity)

? Was gravity pulling on your foil boat? (Yes, gravity is pulling on everything on earth all of the time.)

? Why do you think your boat didn't sink to the bottom of the water?

TRY IT

The idea that water is pushing upward may be new for students. To help make this concept more concrete, do the following demonstration before reading the explanation of buoyancy on page 5.

- Blow up a balloon, tie it off, and let it fall to the floor. *Ask*
  - ? What force caused the balloon to fall? (gravity)
- Next, place the balloon in a tub of water. *Ask*
  - ? Why doesn't the balloon fall to the bottom? What is pushing up on it? (water)
- Submerge the balloon in the water. *Ask*
  - ? What do you think will happen when I let go? (It will pop up.)
- Release the balloon under water. *Ask*
  - ? Why did the balloon pop up? (The water pushed up on it.)

Next, read page 5.

STOP

Ask

? What is the force of the water pushing up on the balloon called? (buoyancy)

DEMONSTRATING BUOYANCY

Explain that in the case of the balloon, the buoyant force of the water was greater than the force of gravity pulling the balloon down, so the balloon was pushed up.

Next, read pages 6–13.

STOP

After reading page 13, show students a 15 cm × 15 cm piece of foil and then crumple it into a ball. *Ask*

? Do you think the foil ball will sink or float?

TRY IT

Place the crumpled-up foil ball in a tub of water. *Ask*

? Why do you think your foil boats float and the foil ball sinks?

? Is one heavier than the other or are they the same weight? (They are the same weight because they are both made of a 15 cm × 15 cm square of the same material.) You may want to weigh each one to prove to students that they are the same weight. It's not uncommon for some students to believe that the weight can change when you change the shape.

Read the text on page 15 that explains that because the foil boat has more volume than the ball of foil, it can displace more water to offset its weight. The more water it displaces, the more the water pushes up on the foil.

Read pages 16 and 17.

STOP

After reading page 17, *ask*

? How does a sinking ship full of too much treasure compare to your boats full of pennies? (When there are too many pennies, the boat starts to sink.)

Read the rest of the book aloud.

elaborate

A Ship for Captain Kidd

Tell students to pretend that Captain Kidd has sent a letter to the class. Read the letter aloud. Tell students that they will be using the materials listed in the letter to design and create a new ship that will hold 50 pieces of "gold" (pennies). Set a time limit (15–20 minutes) for them to design, create, test, and revise their ships. Students will need a new piece of foil and tape each time they revise their design.

As they are working, ask students guiding questions to help them apply the scientific principles they just learned to their ship design. Sample questions:

? What forces are acting on your ship? (Gravity is pulling down and the buoyant force is pushing up.)

Inquiry Place

Have students brainstorm questions about floating and sinking. Examples of such questions include

? Which school supplies will float and which ones will sink? Predict, then test it!

? How can you change something that sinks, like a clay ball, to make it float? Try it!

? How can you change something that floats, like a cork, to make it sink? Try it!

? How does a cruise ship made of tons of steel float on the ocean? Research it!

? Can you create a "flinker"—something that doesn't float or sink, but hangs suspended in the liquid? Try it!

Then have students select a question to investigate or research as a class, or have groups of students vote on the question they want to investigate or research as a team. Have students present this information on a poster. Students will then share their findings in a poster session or gallery walk.

? What can you do to increase the buoyant force on your ship? (Increase the amount of water the ship displaces, or pushes out of the way.)

? What happens if you make the sides too low? (Water floods the ship and it sinks.)

Letter to Captain Kidd

When the time is up, ask each group to share their best ship with the rest of the class. For fun, they can name their ships. *Ask*

? How many different ships did you test?

? Describe the process you used to test your ship to see if it met the challenge. (Students should describe a process where they created, tested, revised, tested again, revised again, and so on.)

? How did your team work together? (Some teams may have split up jobs, some may have had each member try a design, and so on.)

> Connecting to the Common Core
> **Writing**
> Text Types and Purposes: 3.2, 4.2, 5.2

Give each student a copy of the Letter to Captain Kidd student page. Have students write a persuasive letter to Captain Kidd pretending to tell him why he should buy their team's best ship. They should include an explanation of why their ship floated using the words *gravity* and *buoyancy* and a drawing of their ship with arrows showing the directions of the forces of gravity and buoyancy.

More Books to Read

Boothroyd, J. 2011. *What floats? What sinks? A look at density.* Minneapolis, MN: Lerner Publishing.

Summary: This book explains why some things float and others sink and explores the applications of floating and sinking in everyday life.

Branley, F. M. 2007. *Gravity is a mystery*. New York: HarperCollins.

Summary: From the *Let's-Read-and-Find-Out Science* series, this book explains the effects of gravity on Earth and in the solar system.

Bryant-Mole, K. 2002. *Floating and sinking*. Chicago: Heinemann.

Summary: This book describes why things float and sink and presents simple experiments to demonstrate the scientific principles involved.

Name: ______________________________

Float Your Boat

Directions:

1. Use a 15 cm x 15 cm piece of foil to build a boat that can hold some pennies.
2. Now put your boat in the water.
3. Place pennies one by one into your boat.

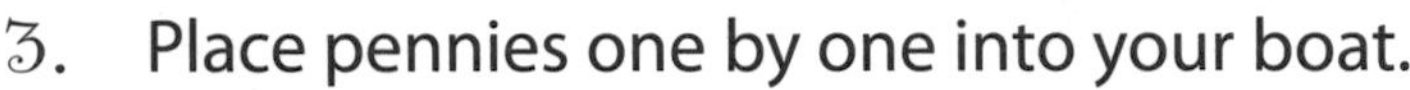

4. In the table below, record the number of pennies each team member's boat held before sinking.

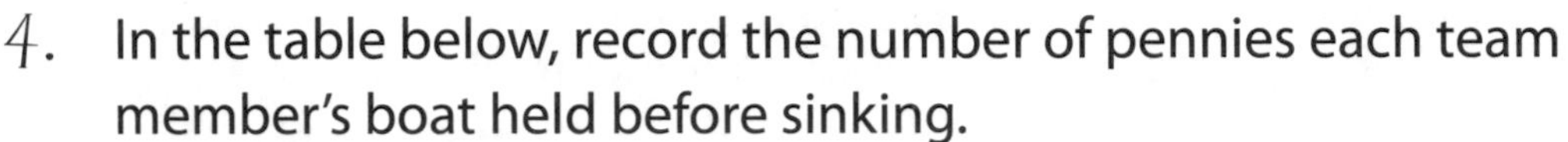

Data Table

Name	Number of Pennies

5. In the box below, draw the boat that held the most pennies.

6. Why do you think that boat held the most pennies before sinking?

Chapter 7

A Ship for Captain Kidd

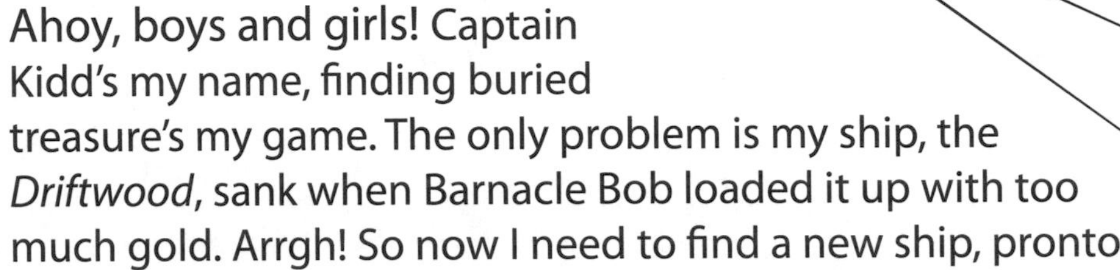

Ahoy, boys and girls! Captain Kidd's my name, finding buried treasure's my game. The only problem is my ship, the *Driftwood*, sank when Barnacle Bob loaded it up with too much gold. Arrgh! So now I need to find a new ship, pronto.

Can you design a ship for me and my crew?

Your ship must:

- Have a sail
- Have a mast to hold up the sail
- Hold 50 pieces of gold (pennies) without sinking
- Be designed and built in less than ___________ minutes

You can use only these materials:

- No more than 30 cm x 30 cm square of heavy-duty aluminum foil
- Scissors
- No more than 50 centimeters of masking tape
- Mast (bendable straw)
- Sail (3″x 5″ index card)
- Markers (for decorating your sail)

Letter to Captain Kidd

Write a letter to persuade Captain Kidd to buy the best ship your team designed. In your letter, be sure to:

- Explain why your ship floats using the terms *gravity* and *buoyancy*
- Include a drawing of your ship
- Draw an arrow showing the direction of the force of gravity on your ship
- Draw an arrow showing the direction of the force of buoyancy on your ship

Dear Captain Kidd,

Drawing

Sincerely yours,

The Wind Blew

Description

What is wind? What is it made of? What can it do? In this lesson, students explore ways to change the speed and direction of a Ping-Pong ball using a handheld air pump to simulate wind. Simple experiments help them understand that air has weight and moving air applies a force to objects. Students investigate how wind strength, opposing wind force, and weight affect the motion of a sailboat.

Suggested Grade Levels: 3-5

LESSON OBJECTIVES *Connecting to the Framework*

PHYSICAL SCIENCES
Core Idea PS1: Matter and Its Interactions
PS1.A: Structure and Properties of Matter
By the end of grade 5: Matter of any type can be subdivided into particles that are too small to see, but even then the matter still exists and can be detected by other means (e.g., by weighing or by its effects on other objects). For example, a model showing that gases are made from matter particles that are too small to see and are moving freely around in space can explain many observations, including the inflation and shape of a balloon [and] the effects of air on larger particles or objects.

Core Idea PS3: Energy
PS3.C: Relationships Between Energy and Forces
By the end of grade 5: When objects collide, the contact forces transfer energy so as to change the objects' motions.

Featured Picture Books 

TITLE: ***The Wind Blew***
AUTHOR: **Pat Hutchins**
ILLUSTRATOR: **Pat Hutchins**
PUBLISHER: **Aladdin**
YEAR: **1993**
GENRE: **Story**
SUMMARY: *Fanciful illustrations and rhyming text describe a day when the wind nearly blew an umbrella, kite, wig, and other items to the sea.*

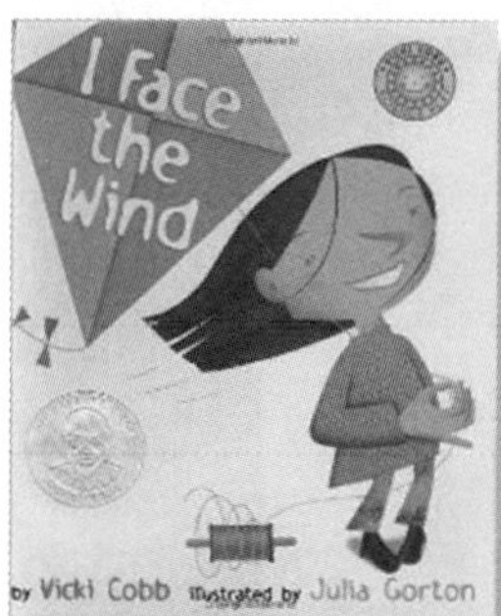

TITLE: ***I Face the Wind***
AUTHOR: **Vicki Cobb**
ILLUSTRATOR: **Julia Gorton**
PUBLISHER: **HarperCollins**
YEAR: **2003**
GENRE: **Non-Narrative Information**
SUMMARY: *Simple, fun experiments help readers discover what wind is made of and why they can feel it.*

Time Needed

This lesson will take about a week. Suggested scheduling is as follows:

Day 1: Engage with *The Wind Blew* Read-Aloud and **Explore** with Wind Challenges

Day 2: Explain with *I Face the Wind* Read-Aloud and experiments

Day 3: Elaborate with *The Wind Blew* Checkpoint Lab, Parts A and B

Day 4: Elaborate with *The Wind Blew* Checkpoint Lab, Parts C and D

Day 5: Evaluate with *The Wind Blew* Checkpoint Lab Part E (Poster Session)

Materials

For Wind Challenges (per pair)

- Plastic handheld air pump
- Ping-Pong ball
- Measuring tape or meterstick
- Cup

> **NOTE:** The Explore portion of this lesson is ideally done on tiled floors. If your classroom floors are carpeted, try to find other areas in your school that have smooth surfaced floors, such as the hallways, gymnasium, or cafeteria.

For* I Face the Wind *Read-Aloud

Per class

- Wire coat hanger
- Pencil
- Two identical balloons or two gallon-size zippered plastic bags
- Tape

Per pair

- Large plastic grocery bag with no holes
- Small ball (a golf ball, racquetball, bouncy ball, or tennis ball will do)

For* The Wind Blew *Checkpoint Lab (per four-person team)

Per class

- Heavy-duty aluminum foil (enough to make boats and to repair them if needed)
- Hole punch
- Sail cut-out and extra sails (in case students need to make repairs)
- Tape for securing sail to straw
- Masking tape for boat name
- Paper towels

Per four-person team

- Wallpaper trough (or other long, shallow container)
- Water for filling trough
- 3 plastic handheld air pumps

- Paper towels
- Soap bar box (or a box of similar size and weight) with the top removed
- Heavy-duty aluminum foil for covering box
- Straw for mast, cut to three-quarters of its length
- Clay for holding mast
- Paper towel
- 15 pennies for weight
- Timer
- Calculator
- Red cup and green cup, taped together
- Team Task Cards (precut)
- Markers
- Small zippered plastic bags to store clay

For Poster Session (per four-person team)

- Poster board or construction paper
- Markers

Student Pages

- Wind Challenges
- The Wind Blew Checkpoint Lab

Background

A Framework for K–12 Science Education suggests that the study of physical science in grades 3–5 be qualitative and conceptual in order to build a foundation for quantitative study in the middle school and high school years. This lesson addresses three core ideas in physical science in a conceptual and integrated manner: matter, energy, and forces and motion. Students learn through activities and reading that wind is moving air and that air is matter, even though you can't see it. Then, by using handheld air pumps to move a Ping-Pong ball, students learn that when matter (like air molecules) collides with another object, the contact force can change the motion of that object. Building on that idea, students experiment with a homemade sailboat to learn that forces have both a strength and direction. Manipulation of various objects develops students' understandings of the forces used to control position and movement.

In this lesson, students use moving air to control the distance and direction a ball travels and investigate how wind force affects the movement of a sailboat. They conduct simple experiments to prove that air is present, even though it can't be seen, and begin to understand that wind is composed of molecules that apply a push when they collide with objects. Their explorations reveal that an object's movement and speed are affected by the direction and strength of a force, as well as the mass of the object itself. Furthermore, when opposing forces are present, the greater of the forces will determine the direction an object travels.

WIND CHALLENGES

The Wind Blew Read-Aloud

Connecting to the Common Core
Reading: Literature
KEY IDEAS AND DETAILS: 3.1, 4.1, 5.1

Inferring

In advance, use paper to hide the cover of the book *The Wind Blew*, and don't tell students the title. Tell them you have a book to share and you want them to infer what is causing the events in the story, without seeing the pictures. When they think they know what "it" is, they can signal (e.g., raise hand or tug ear). When you begin reading aloud, skip the title page and first page of the story, which states, "The wind blew." Instead, begin reading on the next page. ("It took the umbrella from Mr. White and quickly turned it inside out.") Don't reveal the illustrations yet.

Continue reading the rest of the book, stopping after the last page to ask

? What do you think "it" is?

Ask students to whisper their inference to the person next to them. Then ask a few students to share with the entire class. Students will likely have decided the wind caused the events in the story. Reveal the cover of the book and then go back through it to show the illustrations.

Making Connections: Text-to-Self

Ask

? How did you know that the wind caused the events in the story? Could you see it? (Students may say that you can't see the wind, but you can feel it on your skin and see objects moving.)

? Have you ever had an experience with the wind like those in the book? (Have them turn and talk to a partner about their experiences.)

? How could we make our own wind in the classroom? Have several students share their suggestions. (Answers may include a fan, breath, or waving your hand.)

Wind Challenges

Tell students that they are going to be creating their own wind by using a handheld air pump. Give each pair an air pump and have them take turns pumping the air lightly on their partner's hands. Explain that they will be using this method to complete some wind challenges on the floor with a Ping-Pong ball and a partner.

SAFETY

Be careful when working with balls or other equipment on the floor. They can be a serious trip/slip fall hazard. When working with projectiles, students should wear safety glasses or goggles.

Give each pair of students a Ping-Pong ball, measuring tape or meterstick, cup, and the Wind Challenges student page. Encourage teamwork as they complete the challenges on the student page using only the air from the pump:

- Can you make your ball roll more than 1 meter?
- Can you make your ball roll faster?
- Can you make your ball roll straight and then reverse directions?
- Can you make your ball roll into a cup that is lying on its side?
- Can you make your ball roll in a curved path?

Allow students several minutes to work on the challenges. Then bring them back together and *ask*

? How did you make your ball move more than 1 meter? (by positioning the pump behind the ball and pumping fast enough to make it move the distance)

? How did you make your ball roll faster? (by pumping faster)

? How did you make your ball roll straight and then reverse directions? (by pumping the air behind it and then in the opposite direction)

? How did you make your ball roll into a cup? (by positioning the pump directly behind the ball and in line with the cup, and by pumping fast enough to push the ball over the lip of the cup; students may also mention that a partner had to hold the cup in place so that the ball or air did not push it out of position)

? How did you make your ball roll in a curved path? (by pumping short pumps in a curve)

? Did your ball ever go in a direction you didn't mean for it to go? Why do you think this happened? (because the pump was not positioned at the correct angle for making it move in the desired direction)

explain

I Face the Wind Read-Aloud

Connecting to the Common Core
Reading: Informational Text
Key Ideas and Details: 3.3, 4.3, 5.3

Tell students that you have a nonfiction book to help them learn more about wind. Show students the cover of *I Face the Wind* and introduce the author.

> NOTE: Author Vicki Cobb suggests that the best way to use her book is to do the activities described, without rushing, as they come up during the reading. Before you begin reading, make sure you have all the necessary supplies at hand. The author also suggests not turning the page to the explanation until after the child has made the discovery. That way, the book will reinforce what the child has discovered through experience. (See "Note to the Reader" in *I Face the Wind*.)

STOP and TRY IT

While reading aloud *I Face the Wind,* pause where noted to ask questions and try the activities.

First, read aloud pages 1–12.

STOP

After reading page 12, ask

? What is wind made of? (air)

TRY IT

Demonstrate how to trap air inside a plastic grocery bag by pulling the opened bag quickly through the air and twisting the ends closed. Tell the students to try this with a partner. Give each pair a plastic bag with no holes. Caution them not to squeeze their bags hard enough to pop them. Once every student has tried it once, ask the pairs to fill their bags again, twist them closed and hold them. *Ask*

? What will happen if we open the bag? (Some of the air will come out of the bag and into the room.)

Tell everyone to open the bag and hold it open. *Ask*

? If the bag is open, is air still in the bag? (Yes, air is in the bag and in the room, even if we can't see it.)

Collect the bags and continue reading pages 13–18.

SAFETY

When working with plastic bags or balloons, remind students to keep them away from their mouths. These are potential breathing and/or choking hazards.

Trapping Air in a Plastic Bag

STOP

Stop after you read page 18, which asks, "What happens when you hang it [the hanger] on the pencil?" Ask the students to predict what will happen when you hang a hanger holding both the inflated and the noninflated balloons.

TRY IT

Set up the demonstration as instructed by the book with two balloons taped to a hanger. Hook the hanger over a pencil with the two deflated balloons positioned so the hanger is balanced. *Ask*

? Why is the hanger balanced? (The balloons both weigh the same.)

Fill one of the balloons with air, tie it off, and attach it in the same place. *Ask*

? What do you notice about the hanger? (It is tilted, with the inflated balloon hanging lower than the other balloon.)

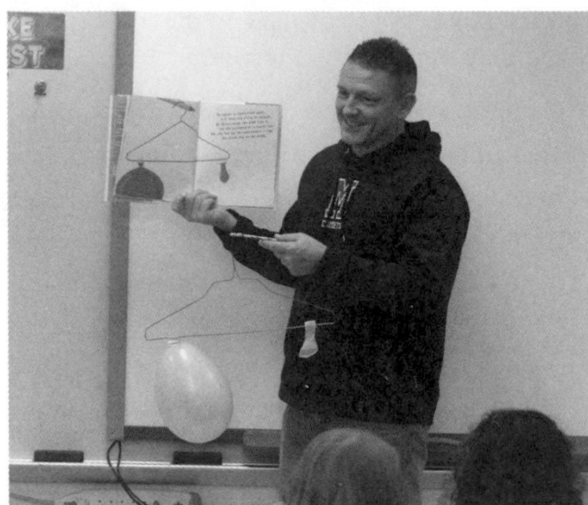

Demonstrating That Air Has Weight

? Why do you think the inflated balloon is hanging lower? (It is heavier because it contains more air.)

? This proves that air has what? (weight or mass)

Continue reading pages 19–24.

STOP

After reading page 24, *ask*

? Which would make a stronger bump, a fast-moving ball or a slow-moving ball?

TRY IT

Give each pair of students a ball and allow them to try the activity described on pages 23 and 24. *Ask*

? Which made a stronger bump, rolling it quickly or slowly? (Rolling it quickly made a stronger bump.)

Explain that in this activity the ball is being used as a model of moving air. The faster the ball moves, the stronger the bump. The faster air molecules move, the more force they are able to apply. A force is a push or pull. You can feel wind because the air molecules are pushing on you.

Read pages 25 and 26.

STOP

After reading page 26, *ask*

? Do you think the wind would be stronger if I wave a book slowly or quickly?

TRY IT

Wave a book slowly and then quickly as you walk around students. *Ask*

? Why is the wind stronger when I wave the book quickly? (The air molecules move faster, causing a stronger push on you, just as rolling the ball more quickly caused a stronger bump.)

Read the rest of the book aloud.

SAFETY

Be careful to quickly wipe up any spilled water on the floor. This is a slip/fall hazard which can result in serious injury.

elaborate

The Wind Blew Checkpoint Lab

In advance, prepare the supplies for the checkpoint lab. See Chapter 3, "Teaching Science Though Inquiry," for a list of tips for managing a checkpoint lab.

1. Cut the tops from the soap bar boxes and reinforce the corners with tape.
2. Cut the straws for the mast to three-quarters of their length. If using bendable straws, cut the straw beneath the bendable section.
3. Tear pieces of foil large enough to cover the box.
4. Fill the wallpaper troughs with water and place them around the room. If you are using a trough with ridges on the inside, be sure to fill the container about three-quarters full to avoid having the boat get stuck.
5. Place the following supplies in a central location for collection and use as needed:
   - Heavy-duty aluminum foil (enough to make boats and to repair them if needed)
   - Hole punch
   - Sail cut-out and extra sails (in case students need to make repairs)
   - Tape for securing sail to straw
   - Piece of masking tape for boat name
   - Paper towels
6. Place the following supplies in a container or bucket for easy collection by each group:
   - 3 air pumps
   - Soap bar box
   - Straw for mast, cut to three-quarters of its length
   - Lump of clay for holding mast
   - Paper towel
   - 15 pennies for weight
   - Timer
   - Calculator
   - Red cup and green cup, taped together
   - Markers
   - Small zippered bags to store clay
7. At the end of each day, have the students remove the clay from their boats and place the clay in a zippered bag to prevent drying.

Tell students that they will be investigating how wind force affects the motion of a boat. Divide students into four-person teams. Give all students a copy of The Wind Blew Checkpoint Lab student page and explain that they will be following the directions to build a sailboat and complete the lab.

Give students the following instructions: "Scientists do the same experiment several times to make sure their results are accurate. You will complete a few trials for each setup. Your results are unlikely to be the same for every trial. Scientists call this variability. You will take an average to determine the time it takes the boat to travel. Ideally, you would repeat the experiments over and over, but due to our time constraints, we will repeat them only three times." Explain how to calculate an average. (Add up the times of all trials, and then divide by the number of trials.)

Give each team member a Team Task Card. Tell students that they may rotate cards on subsequent days.

- The Reader reads the directions as the group is working and is in charge of the green and red cups.
- The Wind Maker uses the pump to create wind to move the boat and chooses a second Wind Maker when necessary.
- The Boat Handler makes changes to the boat, places it in and out of the water, and says "Ready, set, sail!"
- The Materials Manager collects and returns all materials and uses the timer and calculator.

Allow the students to collect their supplies and begin the lab. Check their progress and give assistance when necessary. Each member of the group is responsible for recording data and responses. Before you give a team a check mark or stamp to move forward in the lab, informally evaluate the students by asking probing questions to different members of the team, such as

- How do you know?
- What is your evidence?
- Are you surprised by the results? Why or why not?
- What do you think will happen next?

Redirect their investigations when necessary. As they are working, they should keep the green cup on top of the red cup. If they need help or reach a checkpoint, they should reverse the cups so that the red one is on top.

CHECKPOINT LAB

evaluate

Poster Session

Writing

> Connecting to the Common Core
> **Writing**
> TEXT TYPES AND PURPOSES: 3.2, 4.2, 5.2

Teams that have completed parts A through D of *The Wind Blew* Checkpoint Lab should begin the Poster Session (part E). Explain that scientists and engineers routinely hold poster sessions to communicate their findings to other professionals in their field. Each member of a team will be responsible for creating one section of a poster that communicates the findings from one part of the checkpoint lab (A, B, C, or D). Tell students to use the 3-point rubric in part E to guide their work:

- 3 points for a detailed description of your part of the lab, including a labeled diagram
- 2 points for a conclusion based on evidence
- 1 point for a clear explanation presented to the audience

> Connecting to the Common Core
> **Speaking and Listening**
> PRESENTATION OF KNOWLEDGE AND IDEAS: 3.4, 4.4, 5.4

After all teams have completed their posters, have each team member briefly explain their section of their team's poster. Remind them to speak clearly and at an understandable pace. Encourage the audience to ask questions and provide feedback to the presenters in a polite, respectful manner. Audience members can identify similarities to and differences from their own findings, point out faulty reasoning, and suggest alternative explanations.

After all teams have presented, summarize the concepts the students have investigated in each part of the checkpoint lab. The following explanations of each part of the checkpoint lab are for you. Modify them as necessary to help students understand and apply the concepts.

PART A: Students will discover that the force of their wind moves the boat from one end of the waterway to the other. Greater wind force can move the boat more quickly, but it may also make the boat unsteady, causing it to turn or spin. Students may also discover that the boat slows down as it moves away from them. The sail's ability to "catch" the wind decreases as the air spreads out. Pumping faster will help them compensate for this, as the faster-moving air molecules apply a greater force to the sail. Pumping air toward one side of the sail more than the other will cause the boat to turn. Pumping air on the right will produce a turn to the left, whereas pumping air on the left will produce a turn to the right. Other forces may also contribute to the motion of the boat, such as ridges along the waterway, which may cause the boat to become stuck.

PART B: When two students use the air pumps, more moving air molecules make contact with the sail, pushing it forward more quickly than the force of a single air pump. The boat's speed increases with greater wind force. Some groups may find that the force of two air pumps blowing causes the boat to become unstable. With two Wind Makers, one person may pump faster on a single side of the sail than the other, causing the boat to turn or spin.

PART C: Students are likely to observe that the boat slows down or moves backward when there is an opposing wind force. In the case of two students pumping air in the opposite direction of one, the boat is likely to move in the direction the two air pumps are blowing, at least for a while. When two forces oppose one another, the greater of the two forces will determine the direction of the boat. Distance from the wind force also plays a role. As the boat moves away from one wind force, it is likely to reverse directions as it approaches the opposing wind force.

PART D: If the same force is applied, the speed of a boat will decrease as its weight increases. Heavier objects have greater inertia, a resistance to a change in motion. A greater force is necessary to move these objects. For example, it is easy to move a toy car forward. It takes much greater force to move a real car forward. With the addition of a load, the boat has greater inertia and takes longer to begin moving when a force is applied. For this lesson, it is not important to introduce the concept of inertia unless you desire to do so. The goal is for students to recognize that increased weight affects the boat's ability to move.

After this discussion, give students the opportunity to make changes to their posters before you grade them using the 3-point rubric.

Websites

KidWind Project
www.kidwind.org
Access ideas, lessons, and science fair projects for teaching your students about wind energy.

PBS Kids DragonflyTV "Sailboat"
http://pbskids.org/dragonflytv/show/sailboat.html
Watch a video about two boys who try to discover which sailboat is faster, a double-hull catamaran or a single-hull Lido. The boys build models of their boats to test variables such as sail size, mass, and drag. Then they step inside their boats to have an actual race. Will real life produce the same results?

PBS Kids DragonflyTV "Dragonfly TV Cup"
http://pbskids.org/dragonflytv/games/game_sailing.html
Students play a game in which they adjust the direction and sail position of a virtual boat to navigate a course.

More Books to Read

Branley, F. M. *Air is all around you.* 2006. New York: HarperCollins.

Inquiry Place

Have students brainstorm questions about wind. Examples of such questions include

- ? What causes wind? Research it!
- ? What places on Earth experience more wind than our location? Why? Research it!
- ? How is wind used for energy? Research it!
- ? Does changing the size of a sail affect the motion of a sailboat? Test it!
- ? Does changing the shape of a sail affect the motion of a sailboat? Test it!

Then have students select a question to research or investigate as a class, or have groups of students vote on the question they want to research or investigate as a team. After they make predictions, have them design an experiment to test their predictions. Students can present their findings at a poster session or gallery walk.

Summary: Even though you can't see it, air is everywhere. Interesting facts and simple experiments describe the concept of air and its importance to our world.

Derby, S. *Whoosh went the wind.* 2006. New York: Marshall Cavendish Children.
Summary: A young boy is late for school, and it's all because of the wind! The teacher doubts his outlandish claims, but discovers he's telling the truth when the wind sweeps her out of the classroom window.

Dorros, A. *Feel the wind.* 1990. New York: HarperCollins.
Summary: Simple text and fun facts describe what makes the wind and how it affects the weather. Directions for making a weather vane are also provided.

Name: ______________________________

Wind Challenges

Can you do the challenges below using only your air pump and a Ping-Pong ball?

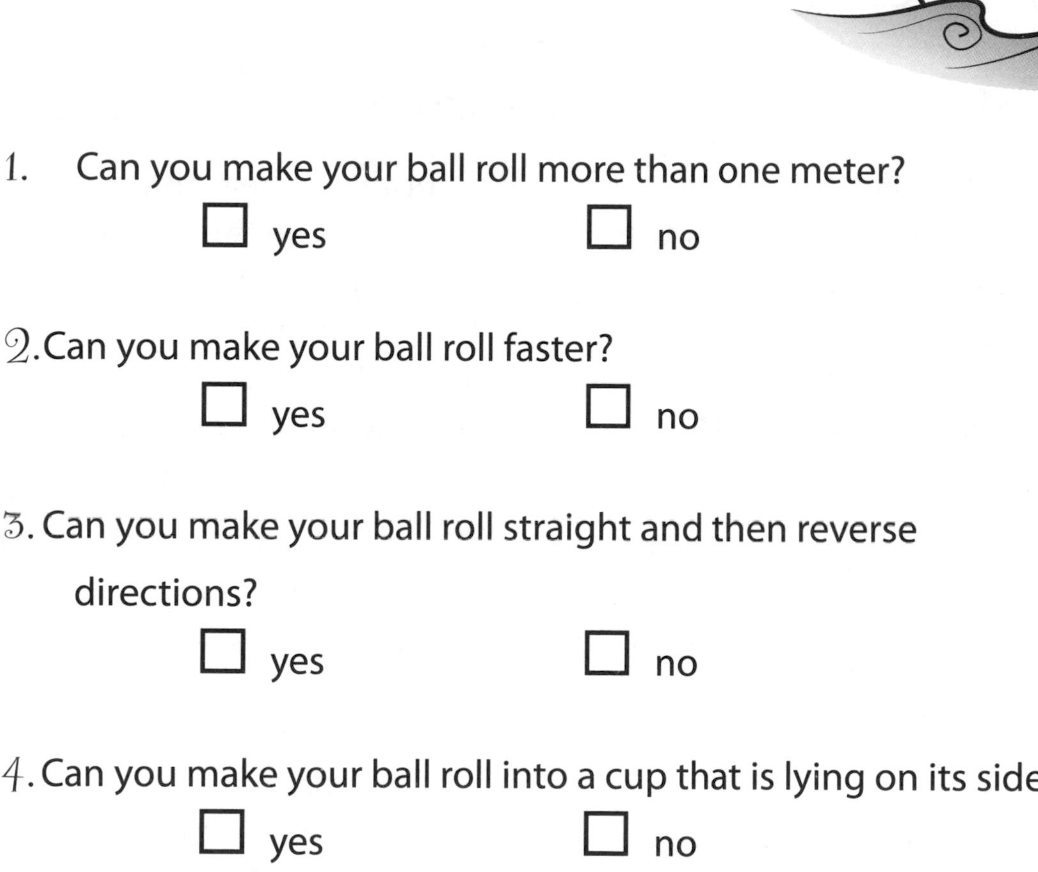

1. Can you make your ball roll more than one meter?

☐ yes ☐ no

2. Can you make your ball roll faster?

☐ yes ☐ no

3. Can you make your ball roll straight and then reverse directions?

☐ yes ☐ no

4. Can you make your ball roll into a cup that is lying on its side?

☐ yes ☐ no

5. Can you make your ball roll in a curved path?

☐ yes ☐ no

The Wind Blew

Team Task Cards

Reader

Read the directions out loud for your team. Put the green cup on top if your group is working. Put the red cup on top if you have a question or if you are ready for a check mark.

Wind Maker

Wait until you hear the directions from the Reader. Then use the air pump to blow on the sail. Keep your pump behind the starting line. Choose a second or third Wind Maker when necessary.

Boat Handler

Wait until you hear the directions from the Reader. Then make changes to the boat if needed. Place the boat in the water. Say, "Ready, set, sail!" Remove the boat when it's not in use.

Materials Manager

Collect all materials. Also use the timer and calculator when needed. When your group receives a check mark, return all materials and have your team help you clean up your workspace.

Name: ______________________________

The Wind Blew

Checkpoint Lab

Follow the directions below. If your team is working, put the green cup on top. If you have a question, put the red cup on top. If you are finished with a part and you are ready for a check from your teacher, put the red cup on top.

Building Your Sailboat

Check the boxes ☑ as your team completes each step.

- ☐ Place the box in the center of your foil. Gently wrap the foil over the sides and into the center of the box. If the foil tears, get a new piece.
- ☐ Place a small lump of clay in the center inside of the boat.
- ☐ Cut out the sail. Use a hole-punch to cut the two circles on the sail.
- ☐ Thread the straw through the two holes in the sail.
- ☐ Put one end of the straw in the clay. Your straw should stand straight up.

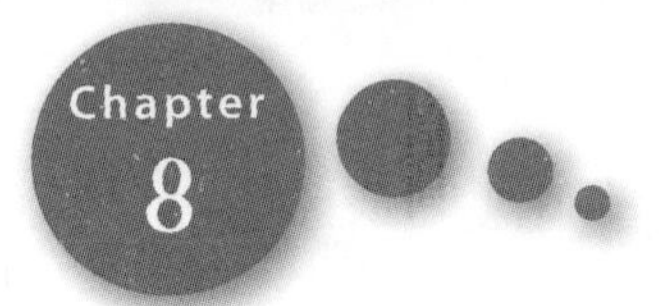

Checkpoint Lab *continued*

The Wind Blew

- [ ] Slide the sail down so that the bottom edge touches the top of the boat.
- [ ] Tape the sail to the straw so that the sail is curved.
- [ ] Think of a name for your boat. Write it on a piece of masking tape and put it on the back of your boat.

How to Protect Your Boat

- If you notice water inside your boat, signal your teacher for help. You may have a leak and need new foil.
- Remove your boat from the waterway and place it on a paper towel when not in use.
- Remove the clay from the straw and boat at the end of every day and place it in a plastic zippered bag. This will keep it from drying out.

Part A

Set Sail!

Throughout this lab, you will be using your air pump to create wind to move the sailboat across a waterway.

- Place the sailboat at one end of the waterway. This will be your starting line.
- Pump air directly toward the boat's sail, but do not allow your pump to pass the starting line at any time. Try to make the boat move to the other end of the waterway.

1. What force caused the sailboat to move across the water?

- Place the sailboat at the starting line.
- This time, pump the air harder on the sail, but do not allow your pump to pass the starting line.

2. Compare the motion of the sailboat to the first time you did this.

- Place the sailboat at the starting line.
- Pump air on the right side of the sail.

Part A *continued*

Set Sail!

3. Which direction does the boat move (right or left)?

____

- Place the sailboat at the starting line again.
- Now pump air on the left side of the sail.

4. Which direction does the boat move (right or left)?

____

5. Write a conclusion. How does the direction of the wind force affect the direction the boat moves?

____

____

____

____

____

____

☐ **Checkpoint A**

Part B

Changing the Amount of Wind

Place the sailboat at the starting line.

- Pump air directly toward the boat's sail, but do not allow your pump to pass the starting line of the waterway.
- Have one team member say, "Ready, set, sail!" On "sail," start the timer as the Wind Maker begins pumping.
- Stop the timer when the sailboat touches the other end of the waterway. Record the time under Trial 1 Time.
- Repeat for Trial 2 and Trial 3.
- Find the average time it took the sailboat to move by adding the three times and dividing by 3.

Number of Wind Makers	Trial 1 Time	Trial 2 Time	Trial 3 Time	Average Time
1				

Make a prediction: If you add another Wind Maker to pump air on the sail, will the boat take a longer or shorter amount of time to move to the other end?

____

____

____

Part B *continued*

Changing the Amount of Wind

- Place the sailboat at the starting line.
- Now add a second Wind Maker so that two people will be pumping air toward the sail.
- Pump directly toward the boat's sail. Do not allow your pumps to pass the starting line of the waterway.
- Have one team member say, "Ready, set, sail!" On "sail," start the timer as the Wind Makers begin pumping.
- Stop the timer when the sailboat touches the other end of the waterway. Record the time under Trial 1 Time.
- Repeat for Trial 2 and Trial 3.
- Find the average time it took the sailboat to move by adding the three times and dividing by 3.

Number of Wind Makers	Trial 1 Time	Trial 2 Time	Trial 3 Time	Average Time
2				

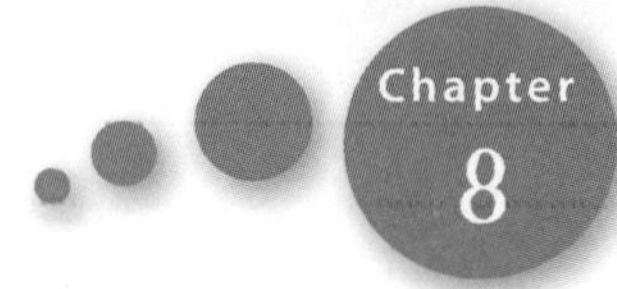

Part B *continued*

Changing the Amount of Wind

1. What was the average time the sailboat moved with one Wind Maker? ______________________

2. What was the average time the sailboat moved with two WindMakers?______________________

3. Did the sailboat take a longer or shorter amount of time to cross the waterway when you added a second Wind Maker?

____

4. Write a conclusion: How does the amount of wind affect the speed of the sailboat?

____

____

____

____

____

____

☐ **Checkpoint B**

Part C

Opposing Winds

- Place the sailboat at the starting line.
- One Wind Maker should sit at the starting line and another Wind Maker should sit at the other end.
- The pumps should not be allowed to pass the end of the waterway at any time.
- Each person will pump directly toward the boat's sail for 10 seconds. The winds will oppose one another.
- Make a prediction: What will happen when both Wind Makers pump air against the sail?______________________________

____

- Have one team member say, "Ready, set, sail!" On "sail," start the timer as the opposing Wind Makers begin pumping air.
- After 10 seconds, say "Stop."

Describe the motion of the sailboat.

____

____

____

____

Explain why you think the sailboat moved this way.

____

____

____

____

Part C *continued*

Opposing Winds

- Now place the sailboat at the starting line again.
- Two Wind Makers should sit at the starting line and one Wind Maker should sit at the other end.
- Each person will pump directly toward the boat's sail for 10 seconds.
- The pumps should not be allowed to pass the end of the waterway at any time.
- Make a prediction: What will happen when two Wind Makers pump air from the starting end of the waterway and one Wind Maker pumps air from the other end?

____

____

____

- Have one team member say, "Ready, set, sail!" On "sail," start the timer as all Wind Makers begin pumping air.
- After 10 seconds, say "Stop."

Describe the motion of the sailboat. ______________________

____

Explain why you think the sailboat moved this way.

____

____

____

____

☐ **Checkpoint C**

Part D

Increasing the Weight of the Sailboat

- Place the sailboat at the starting line.
- Pump air directly toward the boat's sail, but do not allow your pump to pass the starting line of the waterway.
- Have one team member say, "Ready, set, sail!" On "sail," start the timer as the Wind Maker begins pumping.
- Stop the timer when the sailboat touches the other end of the waterway. Record the time under Trial 1 Time.
- Repeat for Trial 2 and Trial 3.
- Find the average time it took the sailboat to move by adding the three times and dividing by 3.

Number of Pennies	Trial 1 Time	Trial 2 Time	Trial 3 Time	Average Time
0				

Make a prediction: If you place 15 pennies inside the boat, will it take a longer or shorter amount of time to move to the other end?

____

____

____

- Now put 15 pennies inside the boat and place it at the starting line.
- Remember to keep your pump behind the starting line at all times.
- Have one team member say, "Ready, set, sail!" On "sail," start the timer as the Wind Maker begins pumping.
- Stop the timer when the sailboat touches the other end of the waterway. Record the time under Trial 1 Time.
- Repeat for Trial 2 and Trial 3.
- Find the average time it took the sailboat to move by adding the three times and dividing by 3.

Number of Pennies	Trial 1 Time	Trial 2 Time	Trial 3 Time	Average Time
15				

1. What was the average time the sailboat moved with zero pennies?____________________

2. What was the average time the sailboat moved with 15 pennies? ____________________

3. Did the sailboat take a longer or shorter amount of time to cross the waterway when you added 15 pennies?

____

4. Write a conclusion: How does the weight of the sailboat affect its speed?

____

____

____

☐ **Checkpoint D**

Part E

Poster Session

Make a poster with your team displaying what you learned about wind forces and motion from The Wind Blew lab. Each member of your team should choose a different part of the lab. Label your section with Part A, Part B, Part C, or Part D.

Here are some things teams should include in each part (A, B, C, and D) of the poster:

3 Points: A detailed description of your part of the lab—A, B, C, or D, including a labeled diagram

2 Points: A conclusion based on evidence (What did you learn about wind forces and motion?)

1 Point: A clear explanation presented to the audience. Be ready to share your poster with the class and answer any questions they might have.

Sail Cutouts

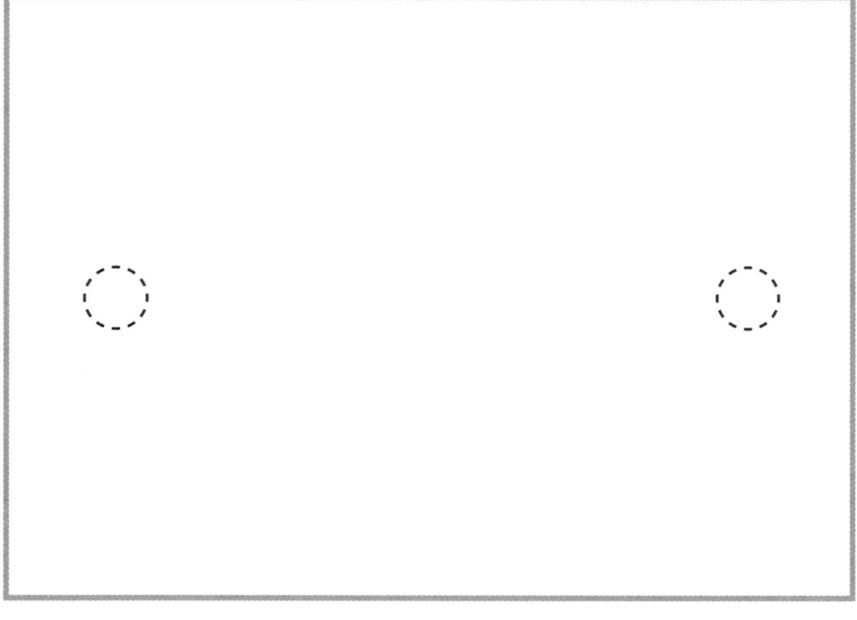

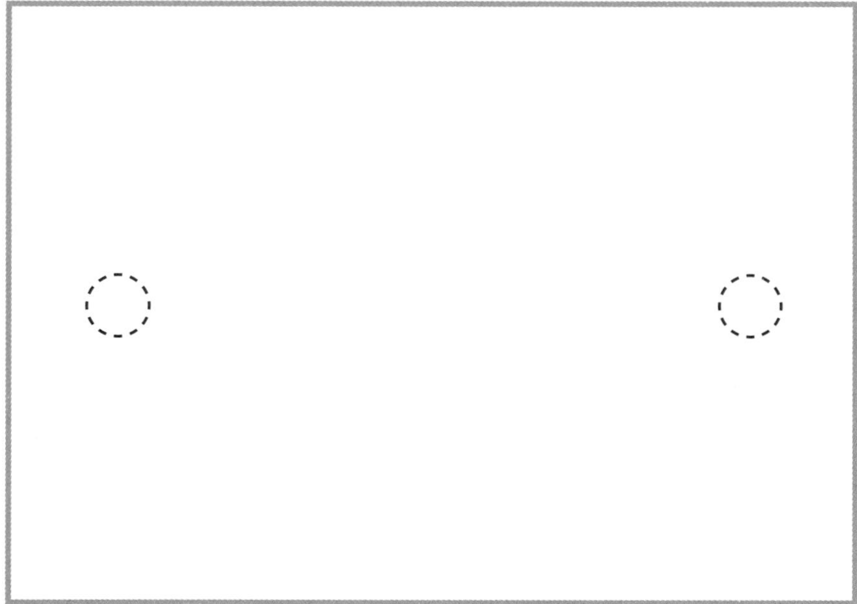

Harnessing the Wind

Description

In this lesson, students are given a real-world context for the concept of energy transfers and transformations through the remarkable true story of a boy who builds a windmill for his village. They learn how a simple generator can transform the energy of motion into electrical energy, and how wind turbines and power plants produce electricity.

Suggested Grade Levels: 3–5

LESSON OBJECTIVES *Connecting to the Framework*

PHYSICAL SCIENCES

Core Idea PS3: Energy

PS3.A: Definitions of Energy

By the end of grade 5: Energy can be moved from place to place by moving objects or through sound, light, or electric currents. (Boundary: At this grade level, no attempt is made to give a precise or complete definition of energy.)

PS3.B: Conservation of Energy and Energy Transfer

By the end of grade 5: Energy can . . . be transferred from place to place by electric currents, which can then be used locally to produce motion, sound, heat, or light. The currents may have been produced to begin with by transforming the energy of motion into electrical energy (e.g., moving water driving a spinning turbine which generates electric currents).

EARTH AND SPACE SCIENCES

Core Idea ESS3: Earth and Human Activity

ESS3.A: Natural Resources

By the end of grade 5: All materials, energy, and fuels that humans use are derived from natural sources, and their use affects the environment in multiple ways. Some resources are renewable over time, and others are not.

Featured Picture Books 

TITLE: *The Boy Who Harnessed the Wind*
AUTHOR: **William Kamkwamba and Bryan Mealer**
ILLUSTRATOR: **Elizabeth Zunon**
PUBLISHER: **Dial Books for Young Readers**
YEAR: **2012**
GENRE: **Narrative Information**
SUMMARY: *Tells the true story of a boy who turned junkyard scraps into a working windmill to bring electricity and water to his famine-struck African village*

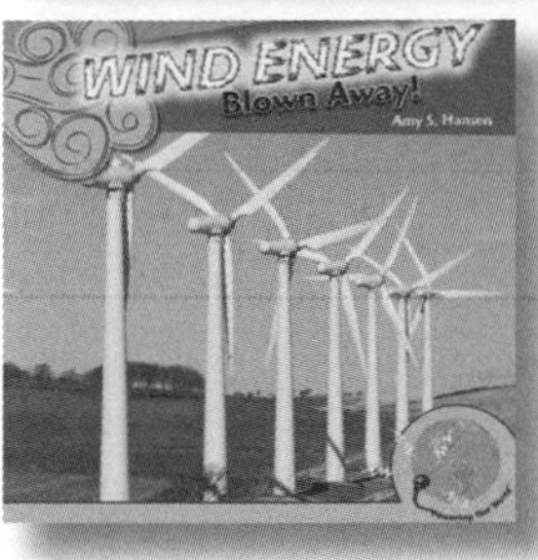

TITLE: *Wind Energy: Blown Away!*
AUTHOR: **Amy S. Hansen**
PUBLISHER: **PowerKids Press**
YEAR: **2010**
GENRE: **Non-Narrative Information**
SUMMARY: *Explains where wind comes from, how wind turbines produce electricity, and the advantages and disadvantages of wind power*

Time Needed

This lesson will take about a week. Suggested scheduling is as follows:

Day 1: **Engage** with *The Boy Who Harnessed the Wind* Read-Aloud and What's in William's Windmill? Preassessment

Day 2: **Explore** with Dynamo Torch, Dynamo Challenge, and Inside the Dynamo

Day 3: **Explain** with How Is the Dynamo Torch Like William's Windmill? T-chart and close reading of "Energy Gets Things Done!" article

Day 4: **Explain** with *Wind Energy: Blown Away!* Read-Aloud and with What's in William's Windmill? Postassessment**,** **Elaborate** with Energy Resources

Day 5: **Evaluate** with Energy Resources Posters

Materials

For The Boy Who Harnessed the Wind Read-Aloud (per class)

- World map or map of Africa
- (Optional) Moving Windmills documentary (from *http://movingwindmills.org/documentary)*

For Dynamo Torch (per group of four to six students)

- 4M Green Science Dynamo Torch kit (from *www.4m-ind.com* or *www.amazon.com*)

For Dynamo Challenge (per group of four to six students)

- Dynamo Torch
- Photos or video of the sOccket (from *http://us.soccket.org*)

For Inside the Dynamo (per group of four to six students)

- Dynamo Torch
- Small Phillips head screwdriver with a #1 bit
- Dynamo Torch kit instruction booklet
- Quart-size zippered plastic bag for Dynamo Torch parts

For How Is the Dynamo Torch Like William's Windmill? (per group of four to six students)

- Disassembled Dynamo Torch
- Dynamo Torch kit instruction booklet

For "Energy Gets Things Done!" Close Reading

- "Energy Gets Things Done!" Close Reading Teacher Page (for teacher use)

For Energy Resources Posters (per group of four to six students)

- Poster board or construction paper
- Markers or colored pencils
- (Optional) Energy Infobooks (from *www.need.org/energy-infobooks*)

Student Pages

- "Energy Gets Things Done!" (article)
- "Energy Gets Things Done!" Close Reading
- Energy Resource Poster Scoring Rubric

Background

Nearly a quarter of the world's population lacks access to any electric power, regardless of how it is produced. It is estimated that 1.5 billion people (concentrated mostly in Africa and southern Asia) live without electricity. This "energy poverty" limits development in the world's poorest places and can make life difficult for people who live there. But one remarkable boy in sub-Saharan Africa helped solve this problem in his own village. William Kamkwamba was born in 1987 in a small town in the nation of Malawi. In 2001 and 2002, his village suffered a devastating famine. William's family went hungry and, with no money to pay for school, he had to drop out. He began visiting a library, started by the American government, where he found books about science. He was particularly intrigued by a picture he saw of a windmill in one of those books. He read that windmills could produce electricity and pump water. His family did not have electricity and he knew that pumping water could help his father, who was a farmer. William was determined to build a windmill. He patched together a tractor fan, shock absorber, frame of a broken bicycle, plastic pipe, bicycle dynamo (a small generator), and some wire to create a windmill that could power a small lightbulb … and he was only 14 years old! He kept tinkering with his design until he was able to use the windmill to charge a car battery, which powered four lightbulbs in his family's home. And several years later, he was able to pump water from a well to water his family's garden.

NOTE: The *Framework* cautions that "the idea that there are different forms of energy, such as thermal energy, mechanical energy, and chemical energy, is misleading, as it implies that the nature of the energy in each of these manifestations is distinct when in fact they all are ultimately, at the atomic scale, some mixture of kinetic energy, stored energy, and radiation. It is likewise misleading to call sound or light a form of energy; they are phenomena that, among their other properties, transfer energy from place to place and between objects" (p. 122). Because many of these descriptive terms associated with forms of energy are arbitrary, we have decided to define some terms when appropriate, while giving examples of other terms to avoid giving students misconceptions. For example, in the article "Energy Gets Things Done!" we are careful to describe energy without actually defining it, because the well-known physical science definition ("the ability to do work") is not meaningful to younger students. Likewise, we do not use the terms *potential* or *kinetic* energy because these appear after the grade 5 endpoint of the *Framework*. We use the term *energy of motion* to describe the energy that is transformed into electrical energy in a generator, and we use the terms *electrical energy* and *electricity* interchangeably. Electrical energy is not defined in terms of the movement of electrons because this concept also appears after the grade 5 endpoint.

In 2007, some journalists found out about William and went to Malawi to see his windmill. William was invited to speak at a conference in Tanzania, and many people were so moved by his story that they donated money to send him back to school. In 2010, William's story became a *New York Times* bestseller for adults, and in 2012 his story was adapted into a children's book of the same title, *The Boy Who Harnessed the Wind*. William went on to study engineering at Dartmouth College.

A Framework for K–12 Science Education suggests that by the end of grade 5 students should understand that electric currents can be produced by transforming the energy of motion into electrical energy, and that energy can be moved from place to place by electric currents. William's remarkable true story, which students learn about on day 1 of this lesson, provides a real-life context for these difficult concepts. His handcrafted windmill was able to transform the energy of motion produced by turning blades into electrical energy to power a lightbulb and other devices for the boy's village. On day 2, students learn how energy can be transformed and transferred by observing how a simple generator, the Dynamo Torch, works. A *generator* is a device that transforms mechanical energy into electrical energy. According to the *Framework*, *mechanical energy* generally refers to some combination of motion and stored energy in an operating machine, and *electrical energy* may mean energy transmitted by electric currents or energy stored in a battery.

Electrical energy can be transferred through wires by currents and used to produce motion, sound, light, or heat. In the 1800s, a scientist named Michael Faraday discovered the remarkable relationship between magnetism and electricity. He figured out that by passing a magnet through a coil of wire or moving a coil of wire near a magnet, he could make an electric current flow through the wire. This process is called *electromagnetic induction* and is the basis for all generators. Inside the Dynamo Torch is a small generator, called a toy motor in the instruction booklet. A motor and a generator are basically the same device, but a motor transforms electrical energy into energy of motion and a generator transforms energy of motion into electrical energy. The generator inside the Dynamo Torch contains a coil of wire and tiny magnets. When the crank is turned, the wire turns between the magnets. This causes the electrons in the wire to flow, producing an electric current through electromagnetic induction. The electric current *transfers*, or carries, electricity through wires and into a tiny LED bulb.

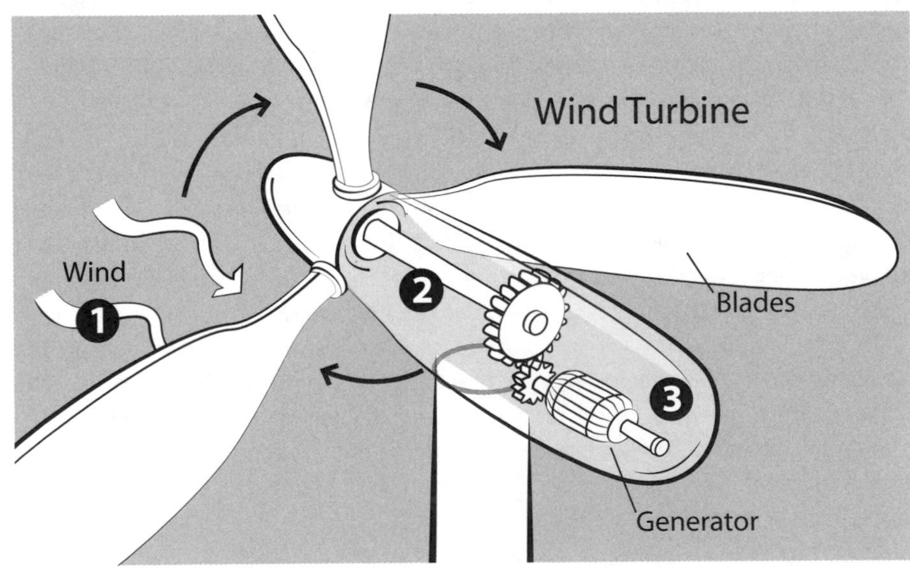

After exploring how the Dynamo Torch works, students do a close reading of an article on day 3 to learn how energy can be both transferred and transformed, what must happen inside the Dynamo Torch to produce electricity, and how energy is transformed to make electricity in power plants. They begin reading "Energy Gets Things Done!" "cold" with no front-loading of vocabulary or prior discussion of content or structure. Next they write a brief summary of the important points the author is trying to convey and share it with a partner. Then they listen closely as the teacher reads the text aloud. (This practice supports the engagement of all students, especially those who struggle with reading the text independently.) After the teacher reads the text aloud, students reread carefully in order to answer a series of text-dependent questions. Here, they must refer back to what the text says explicitly and make logical inferences from it. Finally, students participate in a discussion with a partner to compare the Dynamo Torch with a coal-fired power plant (as opposed to personal reflections on the reading).

On day 4, students listen to a read-aloud of *Wind Energy: Blown Away!* and find out how a *wind turbine* works. In a wind turbine (a windmill that produces electricity), the blades are attached to a short stick called a drive shaft. As shown in the illustration, moving air spins the blades (1), which turn the drive shaft (2). The drive shaft moves gears, which turn the generator (3), which spins magnets around wires to produce electricity. In a wind turbine, the motion of the spinning blades (energy of motion) is transformed, or changed, inside a generator into electrical energy.

All generators, including the ones bringing electricity to our homes, work in the same basic way: they convert energy of motion to electrical energy by rotating magnets around a coil of wire or by spinning a coil of wire in between magnets. The source of the energy of motion may be a wind turbine, water falling through a turbine or waterwheel, compressed air, a hand crank such as the one in the Dynamo Torch, or any other source of energy of motion. Most of the electrical energy consumption in the United States depends on burning fossil fuels, including coal and oil. The heat produced by burning fossil fuels creates steam, which rotates a turbine that spins a generator shaft, which produces electricity through Faraday's principle of electromagnetic induction.

There are a number of problems associated with burning fossil fuels for electricity production, including the creation of greenhouse gases that contribute to global warming and the fact that fossil fuels are nonrenewable and will someday be depleted. In contrast, wind energy is clean, cheap, and renewable. Some disadvantages of wind energy include noise pollution, negative visual impact on the landscape, and the threat that moving blades pose to birds. Nevertheless, wind energy production, though a very small percentage of overall energy production, is on the rise in the United States and worldwide.

The *Framework* states that all energy and fuels that humans use are derived from natural sources, and their use affects the environment in multiple ways. Some resources are renewable over time, and others are not. In the elaborate phase of this lesson, students discuss the advantages and disadvantages of wind energy and burning fossil fuels. Finally, they apply their knowledge of energy transformations and electricity production in the evaluate phase by creating energy resources posters.

engage

The Boy Who Harnessed the Wind Read-Aloud

Connecting to the Common Core
Reading: Literature
Key Ideas and Details: 3.1, 4.1, 5.1

Inferring

Show students the cover of *The Boy Who Harnessed the Wind* and *ask*

? From looking at the cover art and the title, what do you think this book might be about?

Then read the synopsis on the back cover, "The true story of a boy whose great idea and perseverance lit up his home and inspired the world." *Ask*

? What does it mean to persevere? (to keep going in spite of difficulties)

Read the first page and then have students locate Malawi on a map of Africa.

Next, read the story aloud, stopping before the end matter. You may also want to show the award-winning 6-minute documentary *Moving Windmills* (available at *http://movingwindmills.org/documentary*) so students can hear the story in William's own words.

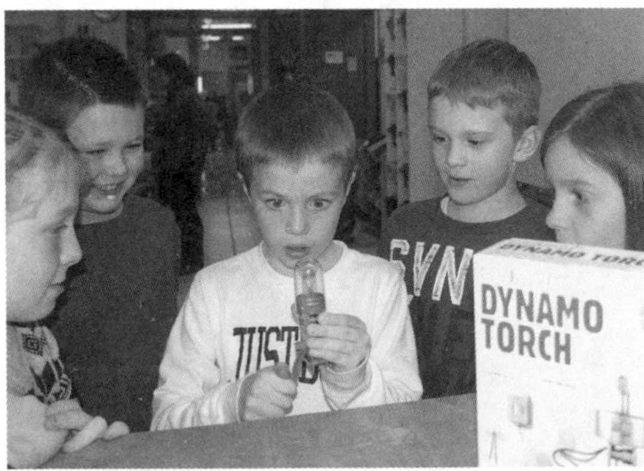

Dynamo Challenge

Questioning

After reading the book, *ask*

? How did William persevere? (His family couldn't afford to send him to school, so he went to the library to learn. People said he was crazy, but he continued to build the windmill.)

Then go back to page 24 where William says, "I have made electric wind" and elicit students' preconceptions by *asking*

? How do you think his windmill could light a lightbulb?

What's in William's Windmill? Preassessment

On a sheet of blank paper, have students write their names and "What's in William's Windmill?" at the top. Then have them draw and describe what they think is inside of William's windmill and how those parts could produce electricity to light a lightbulb. Answers will vary based on students' prior experiences. Collect the papers and tell students they will revisit them later in the lesson.

explore

Dynamo Torch

Tell students that you have a device for them to explore that will help them learn how William's windmill could light a lightbulb. Provide each team of four to six students with a fully assembled Dynamo Torch. Be sure to follow all of the safety guidelines included in the packaging. Allow all students time to explore the device. Ask them to observe what happens to the light when they turn the crank faster and how long the light continues to glow after they stop cranking. Students will discover that the faster they crank, the brighter the light, and as soon as they stop cranking, the light goes out.

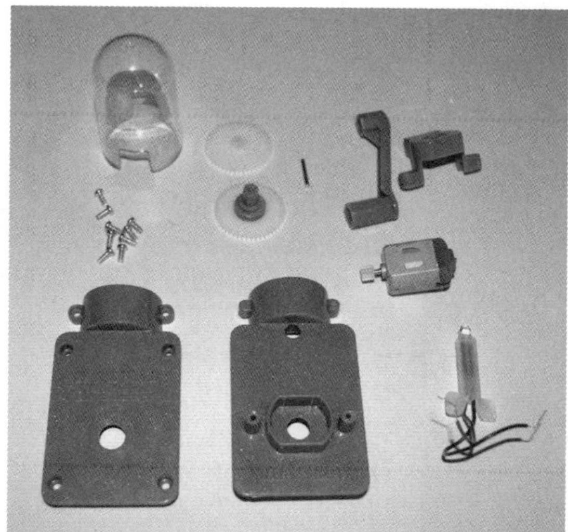

A DISASSEMBLED DYNAMO TORCH

Dynamo Challenge

Have a contest to see which team can keep their lightbulb lit for the longest amount of time. Students will realize very quickly that cranking a device by hand can be very tiring and difficult to sustain. Tell students that this device is similar to one that William used in his windmill, but instead of cranking it by hand, William used wind energy.

Challenge teams to design some other way for the lightbulb to remain lit with less effort. *Ask*

? What else could be used to turn the crank?

Have teams share their ideas (and if time and materials allow, have them build them). Some ideas may be as simple as taking turns cranking or a longer arm for the crank, or as complex as pumping bicycle pedals to turn the crank.

SAFETY

Screwdrivers can cut or puncture skin, so students should wear safety glasses or goggles and use caution when working with screwdrivers.

Tell the students that a group of female students from Harvard University designed a fun way to produce electrical energy using the same scientific principles that the Dynamo Torch uses to produce electricity. Show students some photos or video of the sOccket, a soccer ball that can power a light (*http://us.soccket.org*). The sOccket works in basically the same way as the Dynamo Torch, but instead of using the motion of a hand turning a crank to power a bulb, this ingenious device uses the motion produced from being kicked around. The electricity produced is stored in a battery inside the ball. After playing with the ball for 15 minutes, a small lamp can be plugged into it and produce light for 3 hours. This new invention is actually bringing electricity into places that have not had it before. *Ask*

? What is the secret of the Dynamo and the sOccket ball? What is inside that produces electricity?

? How can we find out? (Take the Dynamo apart to see what is inside.)

Inside the Dynamo

Allow each team of students to carefully disassemble their Dynamo Torch with a small screwdriver; tell them to be sure to keep all of the parts together so that it can be reassembled later. They can refer to the instruction booklet that comes in the kit to identify the parts. Students will discover that the Dynamo Torch contains the following parts:

- Dynamo Torch casing
- Crank
- Transparent torch cover
- Two gears
- LED lamp with holder and wires connected
- Screws
- Toy motor

(Tell students that the toy motor is used as a generator. *Note for teacher:* A motor and a generator are basically the same device. A motor converts electric current to energy of motion, and a generator converts energy of motion to electric current. In the lesson, we refer to it as a generator because we are using it to produce an electric current.)

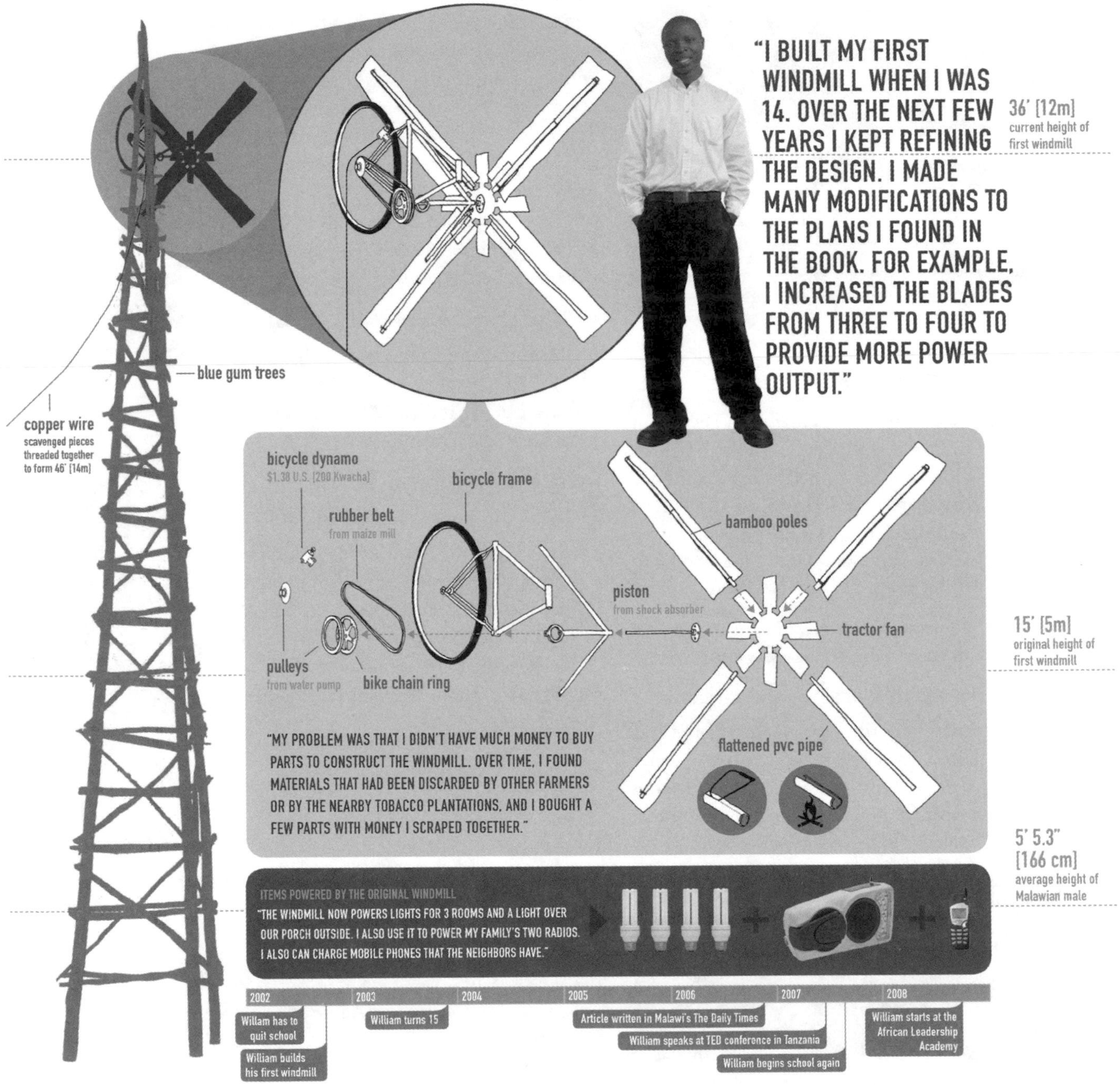

"I BUILT MY FIRST WINDMILL WHEN I WAS 14. OVER THE NEXT FEW YEARS I KEPT REFINING THE DESIGN. I MADE MANY MODIFICATIONS TO THE PLANS I FOUND IN THE BOOK. FOR EXAMPLE, I INCREASED THE BLADES FROM THREE TO FOUR TO PROVIDE MORE POWER OUTPUT."

36' [12m]
current height of first windmill

blue gum trees

copper wire
scavenged pieces threaded together to form 46' [14m]

bicycle dynamo
$1.38 U.S. [200 Kwacha]

bicycle frame

rubber belt
from maize mill

bamboo poles

piston
from shock absorber

tractor fan

15' [5m]
original height of first windmill

pulleys
from water pump

bike chain ring

"MY PROBLEM WAS THAT I DIDN'T HAVE MUCH MONEY TO BUY PARTS TO CONSTRUCT THE WINDMILL. OVER TIME, I FOUND MATERIALS THAT HAD BEEN DISCARDED BY OTHER FARMERS OR BY THE NEARBY TOBACCO PLANTATIONS, AND I BOUGHT A FEW PARTS WITH MONEY I SCRAPED TOGETHER."

flattened pvc pipe

5' 5.3"
[166 cm]
average height of Malawian male

ITEMS POWERED BY THE ORIGINAL WINDMILL
"THE WINDMILL NOW POWERS LIGHTS FOR 3 ROOMS AND A LIGHT OVER OUR PORCH OUTSIDE. I ALSO USE IT TO POWER MY FAMILY'S TWO RADIOS. I ALSO CAN CHARGE MOBILE PHONES THAT THE NEIGHBORS HAVE."

2002 | 2003 | 2004 | 2005 | 2006 | 2007 | 2008

Willam has to quit school

William builds his first windmill

William turns 15

Article written in Malawi's The Daily Times

William speaks at TED conference in Tanzania

William begins school again

William starts at the African Leadership Academy

FIGURE 9.1 WILLIAM'S WINDMILL

As students look at all of the parts, *ask*

? Where do you think the energy that lit the lightbulb came from? (Students will likely infer that the energy was somehow made or produced in the toy motor [actually a generator].)

? What kind of energy makes lightbulbs light up? (electricity or electrical energy)

? Can energy travel? How did the energy from the generator get to the lightbulb? (It traveled through the wires.)

? Can energy change? How did the energy you put into turning the crank change into electricity?

Students should carefully place the parts of the Dynamo Torch into a plastic zippered bag. They will be looking at them again in the "Explain" part of the lesson.

explain

How Is the Dynamo Torch Like William's Windmill?

> Connecting to the Common Core
> **Reading: Informational Text**
> Integration of Knowledge and Ideas: 4.7, 5.7

Project the diagram of William's windmill, which can be found online at *http://movingwindmills.org/story* and is shown in Figure 9.1. Explain that the items William used to build his windmill were copper wire, a bicycle dynamo, pulleys, a bike chain ring, a rubber belt, a piston, a tractor fan, bamboo poles, plastic pipes, a saw, and the frame of a broken bicycle missing a wheel. For blades, he cut the plastic pipes with the saw, melted them over a fire, and flattened them. For a generator, he used the bicycle dynamo, to which he connected the copper wire to a lightbulb in his room.

On the board, draw a T-chart with the headings "Dynamo Torch" (left column) and "William's Windmill" (right column). Write each component of the Dynamo Torch in the left column. Then have students use the diagram of William's windmill to identify each part of the windmill that serves the same purpose. Write those parts in the right column of the T-chart.

Dynamo Torch	William's Windmill
Crank	Blades
Gears	Bicycle Parts
Wires	Copper Wire
Toy Motor *(toy generator)*	Bicycle Dynamo *(a small generator)*

Students should realize that the crank turns like the blades of the windmill, the gears are similar to the bicycle tire and bicycle chain ring, the wires are similar to the copper wire, and the generator (called a toy motor in the instruction booklet of the kit) is similar to the bicycle dynamo that William used. Explain that *dynamo* is another word for a simple generator. *Ask*

? What is inside both the Dynamo Torch and the windmill that somehow produces electricity? (a generator)

? What do you think is inside the sOccket? (a generator)

? What is a generator? How does a generator work?

"Energy Gets Things Done!" Close Reading

> Connecting to the Common Core
> **Reading: Informational Text**
> Key Ideas and Details: 3.1, 4.1, 5.1, 3.2, 4.2, 5.2, 3.3, 4.3, 5.3

Tell students that you have an article that will help them learn the "secret" of how a generator works. Give each student a copy of the "Energy Gets Things Done!" article and the Close Reading student page. Tell them that they are going to do a close reading, which is a careful reading and

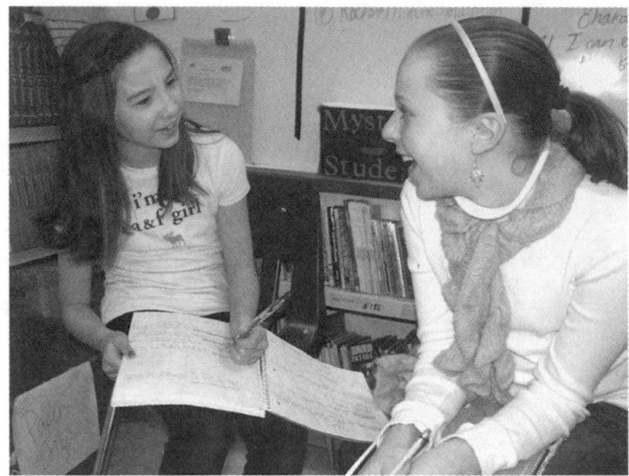

SHARING SUMMARIES

rereading of a text in order to understand it better. If you wish, you can have students complete items 1–4 of the student page in class and assign item 5 for homework. Item 6 can be done in class the following day. See the "Energy Gets Things Done!" Close Reading teacher page for possible correct responses to the items on the student page.

explain

Wind Energy: Blown Away! Read-Aloud

Hand back the papers students created on day 1 with their ideas about what was inside William's windmill. Tell them they can learn more about what is inside a windmill from the book *Wind Energy: Blown Away!* Show students the cover of the book and explain that the cover photo shows a wind farm, a place where electricity is produced by huge windmills called wind turbines.

Using Features of Nonfiction

Connecting to the Common Core
Reading: Informational Text
CRAFT AND STRUCTURE: 4.5

Show students the table of contents in *Wind Energy* and explain that with a nonfiction book you do not have to read it from cover to cover like a story. You can enter the text at any point and only read the information you need. In this case, you want information about how wind turbines produce electricity. Read through the section titles in the table of contents and *ask*

? Which section should we read to help answer our question about how windmills produce electricity? ("Electricity From Wind Turbines," page 10)

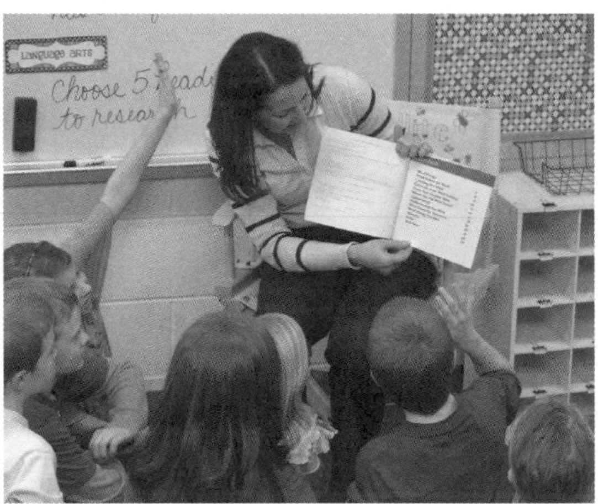

USING THE TABLE OF CONTENTS

Questioning

Connecting to the Common Core
Reading: Informational Text
KEY IDEAS AND DETAILS: 4.1, 5.1

As you read page 10 aloud, have students listen to the explanation of how wind turbines produce electricity. Stop and discuss the explanation.

From the reading, students will learn that the electricity in a wind turbine is generated by moving magnets around coils of wire, similar to the Dynamo Torch. (However, in the Dynamo Torch the wire is spinning and the magnets are fixed. Remind students that Michael Faraday discovered that electric current can be produced both ways:

spinning magnets within coils of wire or spinning coils of wires between magnets.) The blades of a wind turbine are attached to a stick called a drive shaft. When the wind turns the blades, the blades turn the drive shaft, which then moves gears. The gears turn the generator, which spins magnets around wires, and this makes electric current.

Ask

- ? Wind turns the blades and shaft of the windmill. What kind of energy is being used to move the blades? (energy of motion)
- ? The turning shaft moves gears, which make a magnet spin around a coil of wire. What kind of energy does this generate? (electrical energy)
- ? How do you think the electrical energy gets to where it is used? (It is transferred by electric currents through wires.)

What's in William's Windmill? Postassessment

Writing

> Connecting to the Common Core
> **Writing**
> TEXT TYPES AND PURPOSES: 4.2, 5.2

After discussing page 10 of Wind Energy, have students revise the papers they wrote on day 1, where they guessed what might be inside William's windmill. Their revised explanations and drawings should reflect what they have learned from their Dynamo Torch explorations, the "Energy Gets Things Done!" article, and page 10 of Wind Energy. Encourage them to include the words *energy of motion, electrical energy, transform,* and *transfer* in their explanations. They should be able to explain how the energy of motion from the wind turning the blades of the windmill is transformed into electrical energy inside a generator (or bicycle dynamo, in the case of William's windmill), which is then transferred through wires to a lightbulb where it produces light.

elaborate

Energy Resources

> Connecting to the Common Core
> **Speaking and Listening**
> COMPREHENSION AND COLLABORATION: 3.1, 4.1, 5.1, 3.2, 4.2

Ask

- ? What do you think are the advantages of using wind energy? (Wind never runs out. Some students will realize that using wind energy does not cause air pollution.)

Tell students that it is important to understand, as citizens of the planet, that there are both advantages and disadvantages to every source of electricity. All materials, energy, and fuels that humans use come from nature, and their use affects humans and the environment in multiple ways.

Ask

- ? Can you recall from the article you read, "Energy Gets Things Done!" where most of the electricity used in the United States comes from? (from burning fossil fuels)

Next, read pages 4 and 20 of *Wind Energy: Blown Away!*, which explain that most of the electricity used in the United States comes from fossil fuels, that wind energy is renewable and nonpolluting, that the cost of it has been going down, and that more and more wind farms are being constructed. Then discuss the terms *renewable* and *nonrenewable*. *Ask*

- ? What are the disadvantages of burning fossil fuels, like coal? (Answers will vary, but some students will know that fossil fuels are nonrenewable and that burning them produces air pollution.)

Explain that some energy resources, like wind, are renewable over time, while others, like fossil fuels (including coal and oil), are not.

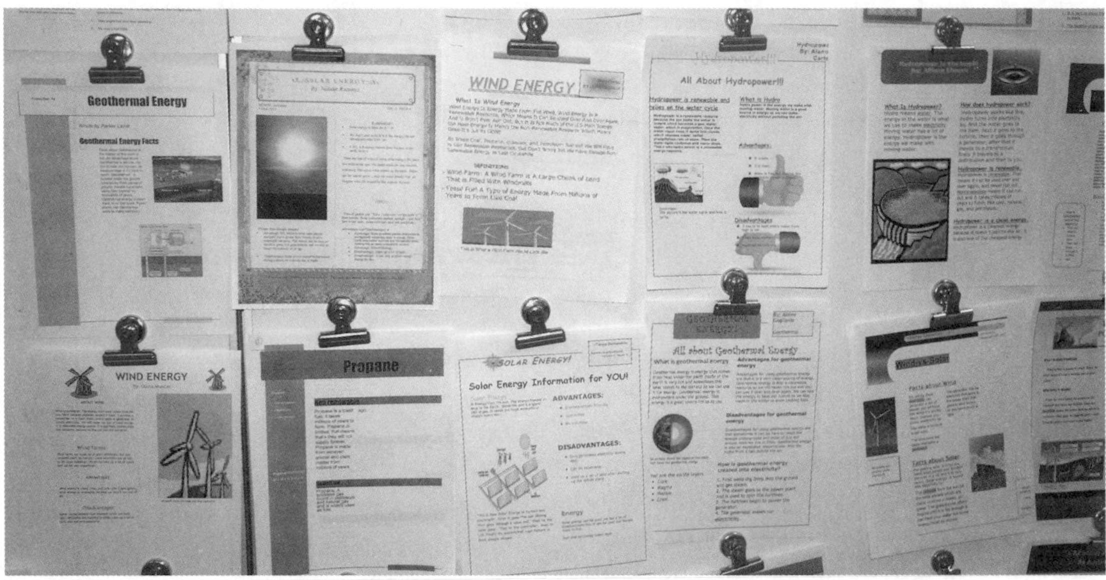

ENERGY RESOURCE PAMPHLETS

Ask

? What do you think are the disadvantages of using wind energy? (Answers will vary, but some students may think that the turbines are visually unappealing or infer that when the wind doesn't blow no electricity is produced.)

Then read page 18 of *Wind Energy* and discuss the disadvantages of wind energy. *Ask*

? Would you want to put wind turbines on your property? Would you want to live near a wind farm? Would you want to get your electricity from a wind farm? Why or why not?

Energy Resources Posters

Writing

Remind students that it is important to understand that there are advantages and disadvantages to every energy resource. All energy resources that we use to produce electricity come from nature, and their use affects humans and the environment in multiple ways. Some resources are renewable over time, and others are not. Then tell students

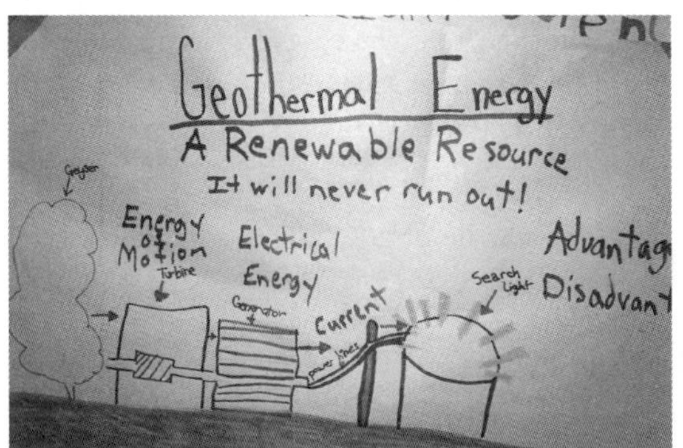

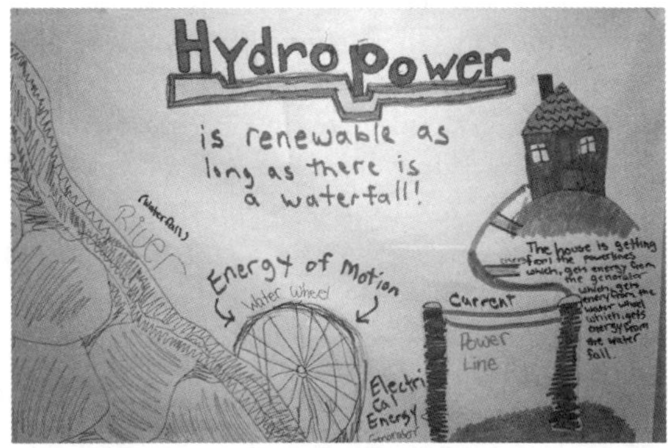

ENERGY RESOURCE POSTERS

Inquiry Place

Have students brainstorm questions about magnetism and electricity. Examples of such questions include

? Can you build an electromagnet? Try it! (Simple instructions can be found at *www.sciencebob.com/experiments/electromagnet.php*; be sure to follow all safety precautions)

? Does the thickness or length of the nail affect the strength of an electromagnet? Test it!

? Does the number of times you wrap the wire around the nail affect the strength of an electromagnet? Test it!

? Can you build a simple Faraday motor? Try it! (Instructions can be found at *www.youtube.com/watch?v=k7JTyRBfeF4&feature=player_embedded#*; be sure to follow all safety precautions.)

? Can you build a simple electric generator? Try it! (Instructions can be found at *www.amasci.com/amateur/coilgen.html*; be sure to follow all safety precautions.)

Then have students select a question to investigate or a device to build, or have groups of students vote on the question they want to investigate or the device they want to build as a team. Students can present their findings at a poster session or gallery walk.

that they will be working in teams to research an energy resource and create a poster about the resource to the class. (*Note:* You may want to have students work individually to produce a pamphlet about an energy resource, rather than having teams create a poster.) Energy resources to choose from should include hydropower, biomass, geothermal, solar, uranium (nuclear), petroleum, natural gas, and coal. The National Energy Education Development (NEED) Project has excellent online materials called Energy Infobooks that students can download to research their energy resource (*www.need.org/Energy-infobooks*).

Connecting to the Common Core
Writing
TEXT TYPES AND PURPOSES: 4.2, 5.2

As shown on the Energy Resource Poster Scoring Rubric, posters should include the following elements:

- The name of the resource, how it is collected, where it is used, and why it is renewable or nonrenewable
- A description of how it is used to produce electricity, including the terms *transform* and *transfer*
- A labeled diagram showing how it is used to produce electricity
- The advantages and disadvantages of its use, including any negative impacts on humans and the environment
- Information on new technologies that are being developed to improve it, or current news about it

Additionally, students should write and perform a skit, song, or rap to "sell it" to consumers!

Students can present their posters in a poster session or gallery walk.

Websites

NEED Project Energy Infobooks
http://www.need.org/energy-infobooks

sOccket
http://us.soccket.org

William Kamkwamba's story (six-minute documentary)
http://movingwindmills.org/documentary

William Kamkwamba's story and windmill diagram
http://movingwindmills.org/story

More Books to Read

Drummond, A. 2011. *Energy Island: How one community harnessed the wind and changed their world.* New York: Farrar, Straus and Giroux.
Summary: At a time when most countries are producing ever-increasing amounts of carbon dioxide, the rather ordinary citizens of Samsø, Denmark, have accomplished something extraordinary—in just 10 years they have reduced their carbon emissions by 140% and become almost completely energy independent. A narrative tale and a science book in one, this inspiring true story proves that with a little hard work and a big idea, anyone can make a huge step toward energy conservation.

Enz, T. 2012. *Harness it: Invent new ways to harness energy and nature.* Mankato, MN: Capstone Press.
Summary: Written by a mechanical engineer, this book explains the principles of inventing and provides photo-illustrated instructions for making a variety of devices to harness energy, including a solar-powered marshmallow roaster and a wind turbine.

Hansen, A. 2010. *Hydropower: Making a splash!* New York: PowerKids Press.
Summary: This book from the *Powering Our World* series describes how fast-flowing water from waterfalls or dams can be harnessed to generate electricity. The author discusses the advantages and disadvantages of hydropower, including the fact that dams can cause flooding and destroy fish and animal habitats. Other titles in the series include *Fossil Fuels: Buried in the Earth, Geothermal Energy: Hot Stuff! Nuclear Energy: Amazing Atoms, Solar Energy: Running on Sunshine,* and *Wind Energy: Blown Away!*

Leedy, L. 2011. *The shocking truth about energy.* New York: Holiday House.
Summary: This highly engaging book is packed full of information about many aspects of energy, including different forms of energy, energy transformations, and energy sources.

Seuling, B. 2003. *Flick a switch: How electricity gets to your home.* New York: Holiday House.
Summary: This book describes how electricity was discovered, how early devices were invented to make use of it, how it is generated in power plants, and how it is distributed for many different uses.

Energy Gets Things Done!

What Is Energy?

Think about how you traveled to school today. Did you walk? Pedal your bike? Ride in a bus or car? When you got to school, did you sharpen a pencil, open a book, use a computer, or flip a switch to turn on a light? All of these things required energy. **Energy** is involved in everything that happens, everywhere it happens. Energy makes things light up or heat up. It makes things move or get loud. It makes us grow and change. Energy gets things done!

Energy Can Move

Energy does not usually stay in one form or remain in one place. Energy can be transferred, or moved, from one location or object to another. For example, energy can be **transferred** through wires by **electric currents,** which can produce motion, sound, heat, or light, depending upon the type of device connected to the circuit. When you turn on a light in your home, you are using electric energy that has most likely been transferred through wires for MILES before it reaches the lightbulb!

Energy Can Change

Energy can also be **transformed,** or changed, from one form to another. You can see how energy transforms by using a simple **generator** such as a Dynamo. A generator is a machine that changes the **energy of motion** into **electrical energy.** To get the bulb in the Dynamo to light, you have to provide energy by turning the crank with your hand. The crank turns two gears connected to a coil of wire located between tiny magnets. When the coil of wire spins between the magnets, an electric current is produced. Energy of motion has been *transformed* into electrical energy! The electric current then *transfers* electrical energy through wires into a lightbulb where it produces light (and just a tiny bit of heat as well).

Dynamo Torch

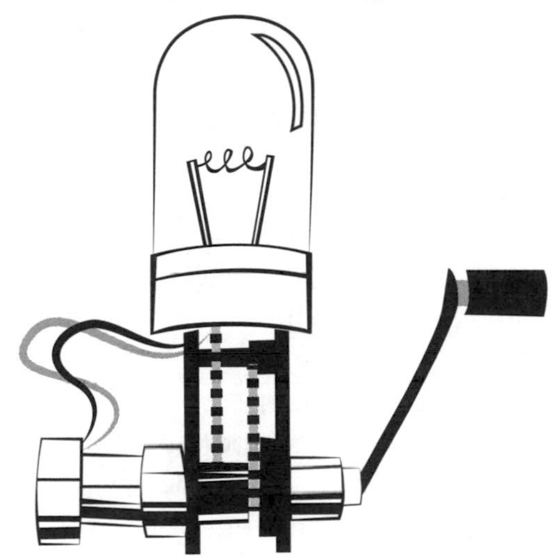

Electromagnetic Induction

A scientist named Michael Faraday invented the first generator in the early 1800s. He figured out that moving a coil of wire between magnets, or moving a magnet through a coil of wire produced electric current. This process, called **electromagnetic induction,** is the basis for all generators. So when you turn on a light in your home, you are using electrical energy that has been produced by a generator.

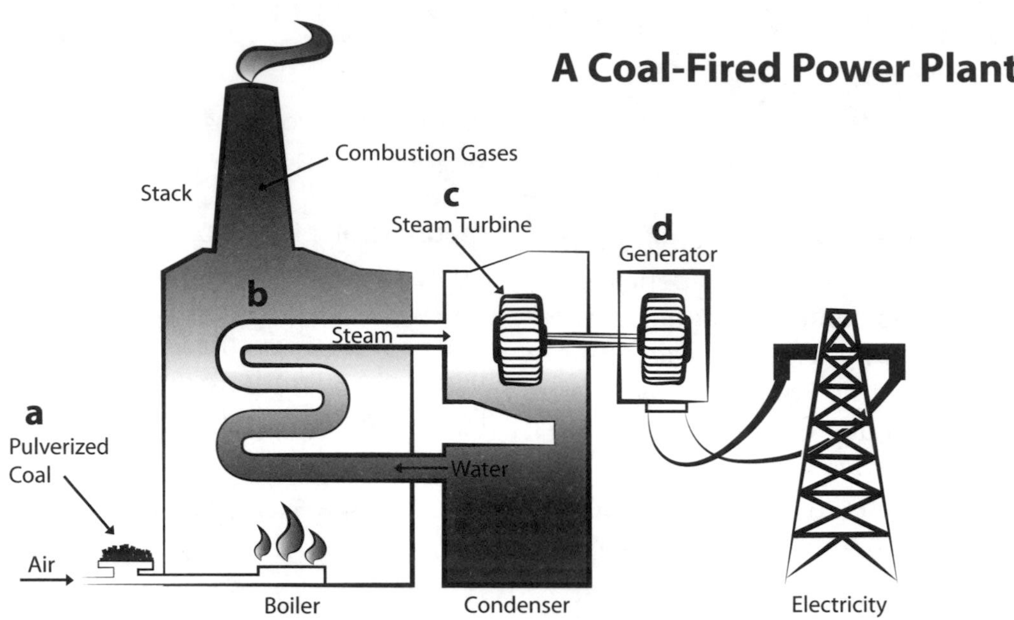

How Electricity Gets to Your Home

Many things in your home depend upon electricity to work, from lights to kitchen appliances to computers. So where does that electricity come from? It depends upon what kind of power plant your community depends upon for electricity. A **power plant** is a place where a large generator transforms energy of motion into electrical energy. If you live near a dam, your electricity might be produced in a **hydroelectric** power plant. Here, the energy to turn **turbines** (machines with blades attached to a shaft) that turn a generator comes from the force of falling water. If you live in a windy place, your electricity might be produced in a **wind farm**. Here, the energy to turn the wind turbines that turn the generators inside them comes from the wind. In most places in the United States, however, electricity is produced by burning **fossil fuels**, as in the **coal-fired power plant** in the diagram above.

First, pulverized, or finely powdered coal **(a)** is burned to boil the water inside a boiler **(b)**. Steam builds up and enormous pressure forces the blades of a steam turbine **(c)** to spin. The spinning turbine is connected to the shaft of a generator **(d)**, which turns to spin magnets within wire coils. This interaction between the magnets and the coils of wire is what produces electrical energy, just like in the Dynamo! To sum up how electricity gets to most homes: Energy produced by burning coal, falling water, wind, or other sources turns a turbine. The turbine turns a generator, which transforms the energy of motion into electrical energy. The electrical energy is then transferred by electric currents through a network of wires and transformers until it reaches your home. So the next time you flip a switch, think about the remarkable chain of events that has to occur in order for that light to go on!

Name: ____________________

Energy Gets Things Done!

Close Reading

1. Read the article "Energy Gets Things Done!" silently to yourself.

2. Write a brief summary of the article, highlighting the most important points the author is making.

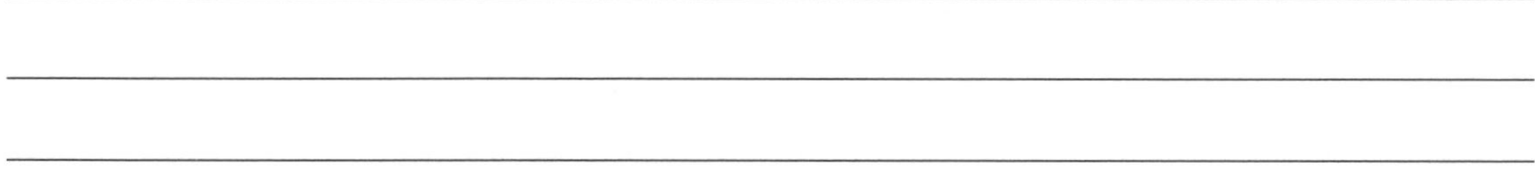

3. Share your summary with a partner. Discuss how your summaries are the same and how they are different.

4. Listen closely as your teacher reads the article aloud.

Energy Gets Things Done! (Continued)

5. Reread the article to answer the following questions:

   a. What is an example from the article of how energy can be transferred from one place to another?

   b. What is an example from the article of how energy can be transformed from one form to another?

   c. What sequence of events must occur to light the bulb in a Dynamo Torch?

   d. Why do you think the author included information about Michael Faraday?

   e. What do generators do?

   f. Where does the energy come from to turn the turbines in a hydroelectric power plant?

6. Reread the section entitled, "How Electricity Gets to Your Home." Then discuss the following questions with a partner:

   How is a Dynamo Torch like a coal-fired power plant? How is it different?

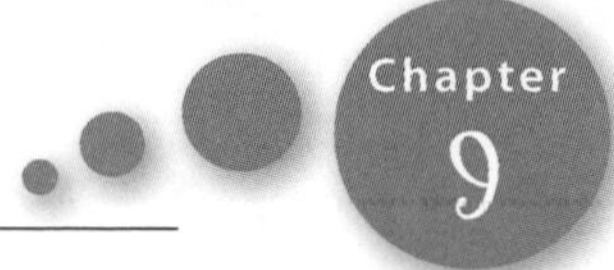

Name: ______________________________

Energy Resource Poster

Scoring Rubric

Research an energy resource used to produce electricity. Include the criteria listed below on your poster.

4-Excellent 3-Above Average 2-Average 1-Below Average

Score	Criteria
____ 4 ____ 3 ____ 2 ____ 1	The name of the resource, how it is collected, where it is used, and why it is renewable or nonrenewable
____ 4 ____ 3 ____ 2 ____ 1	A description of how it is used to produce electricity, including the terms *transform* and *transfer*
____ 4 ____ 3 ____ 2 ____ 1	A labeled diagram showing how it is used to produce electricity
____ 4 ____ 3 ____ 2 ____ 1	The advantages and disadvantages of its use, including any negative impacts on humans and the environment
____ 4 ____ 3 ____ 2 ____ 1	Information on new technologies that are being developed to improve it, or current news about it
____ 4 ____ 3 ____ 2 ____ 1	Write and perform a skit, song, or rap to "sell it" to consumers!

____ **Total Points/24**

Energy Gets Things Done!

Close Reading

The Close Reading student page instructions are listed below, with possible correct responses in parentheses.

1. Read the article "Energy Gets Things Done!" silently to yourself.

2. Write a brief summary of the article, highlighting the most important points the author is making. (Energy is involved in everything that happens. It can move and change forms. A dynamo is a generator that transforms the energy of motion from your hand turning the crank into electrical energy. All generators work by electromagnetic induction, discovered by Michael Faraday. Most homes get their electricity from power plants that burn fossil fuels to turn steam turbines that turn generators. A lot of things have to happen to get electricity into our homes.)

3. Share your summary with a partner. Discuss how your summaries are the same and how they are different.

4. Listen closely as your teacher reads the article aloud.

5. Reread the article to answer the following questions:

a. What is an example from the article of how energy can be transferred from one place to another? (It can be transferred through wires by electric currents.)

b. What is an example from the article of how energy can be transformed from one form to another? (Energy of motion can be transformed into electrical energy inside a generator.)

c. What sequence of events must occur in order to light the bulb in a Dynamo Torch? (Your hand turns the crank. The crank turns gears connected to a coil of wire located between tiny magnets. When the coil of wire spins between the magnets, an electric current is produced. The electric current transfers electrical energy through wires to light the bulb.)

d. Why do you think the author included information about Michael Faraday? (The author thought it was important for the reader to know that electromagnetic induction is the basis for all generators, and that Michael Faraday was the scientist who discovered this.)

e. What do generators do? (They transform energy of motion into electrical energy.)

f. Where does the energy come from to turn the turbines in a hydroelectric power plant? (It comes from falling water.)

6. Reread the section entitled, "How Electricity Gets to Your Home."

Then discuss the following questions with a partner:

How is a Dynamo Torch like a coal-fired power plant? How is it different? (Discussions may include the following points: A Dynamo Torch and a coal-fired power plant both produce electricity, they both contain generators, they both change energy of motion into electrical energy, etc. A Dynamo Torch is much smaller than a power plant and only produces a small amount of electricity. A Dynamo Torch gets energy of motion from your hand turning a crank, whereas a coal-fired power plant gets energy of motion from burning coal to produce the steam that spins the turbine. Students may realize that a Dynamo Torch does not produce air pollution, but a coal-fired power plant does.)

Sounds All Around

Description

In this lesson, students discover that sound is caused by vibrating matter and that sound can make matter vibrate. They also construct a string telephone to explore how sound can travel through a solid object.

Suggested Grade Levels: K–2

LESSON OBJECTIVES *Connecting to the Framework*

PHYSICAL SCIENCES

Core Idea PS4: Waves and Their Applications in Technologies for Information Transfer
PS4A: Wave Properties
By the end of grade 2: Sound can make matter vibrate, and vibrating matter can make sound.

Featured Picture Books 

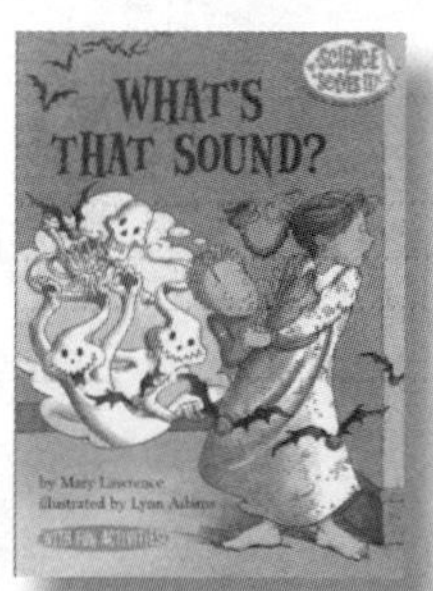

TITLE: ***What's That Sound?***
AUTHOR: **Mary Lawrence**
ILLUSTRSTOR: **Lynn Adams**
PUBLISHER: **Kane Press**
YEAR: **2002**
GENRE: **Dual Purpose**
SUMMARY: *Tim hears strange noises that keep him up at night when his family stays in a country cottage.*

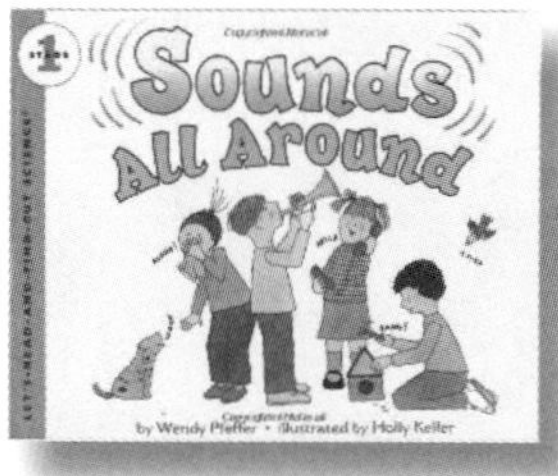

TITLE: ***Sounds All Around***
AUTHOR: **Wendy Pfeffer**
ILLUSTRATOR: **Holly Keller**
PUBLISHER: **HarperCollins**
YEAR: **1999**
GENRE: **Non-Narrative Information**
SUMMARY: *Describes sounds in the human and natural worlds, how sound travels, and how it can vary.*

Time Needed

This lesson will take several class periods. Suggested scheduling is as follows:

Day 1: **Engage** with Sound Anticipation Guide ("Before") and *What's That Sound?* Read-Aloud, and **Explore** with Listening Activity

Day 2: **Explore** with Sound Stations and **Explain** with Sound Stations Discussion

Day 3: **Explain** with *Sounds All Around* Read-Aloud and **Evaluate** with Sound Anticipation Guide ("After")

Day 4: **Elaborate** and **Evaluate** with Phone a Friend

Materials

For Sound Stations

Station 1: Feel the Beat

- Drum
- Mallet
- Teaspoon of dry rice
- Bowl
- Plastic wrap
- Tape

Station 2: Sound Hanger

- Wire hanger
- String
- Desk

Station 3: Underwater

- Large bucket of water
- 2 rocks, each about the size of a golf ball
- Paper towels

Station 4: Cardboard Kazoo

- Cardboard toilet paper or paper towel tubes
- Markers
- Tissue paper squares, 3 × 3 inches
- Rubber bands

Station 5: Knock, Knock

- Desk or table

SAFETY

- Tell students never to eat or taste food (e.g., rice) or a beverage that has been made or used in the lab or a classroom activity unless instructed to do so by the teacher.
- Be careful to quickly wipe up any spilled water, oil, or other liquid on the floor. This is a slip/fall hazard, which can result in a serious injury.
- Any activity equipment placed in the ear (e.g., a funnel) needs to be sanitized before being used by another student.

Station 6: The Doctor Is In

- Funnel
- 50 cm of plastic aquarium tubing to fit the narrow part of the funnel
- Tape

Station 7: Mystery Sounds

- Six opaque film canisters or plastic Easter eggs labeled A–F and filled with common objects, such as rice, pennies, paper clips, jingle bells, dry cereal, popcorn kernels

For Phone a Friend (per pair)

- 3 m cotton string
- 2 plastic or paper cups with holes in the bottom
- 2 paper clips

Student Pages

- Sound Anticipation Guide
- Sound Stations
- Phone a Friend

Background

Sound is a type of energy made by vibrating matter. The vibrations travel away from their source in waves. Sound waves can travel through all types of matter: solids, liquids, and gases. Anything that sound travels through is called a medium. When something vibrates, it causes movement in the tiny particles that make up the medium. These particles bump into the particles next to them, which makes them bump into more particles. Sound waves can cause matter to vibrate, for example, sound waves from thunder can travel through the air and make windows rattle. Sound cannot travel through a vacuum or outer space because neither contains a medium for vibrations to travel through. *A Framework for K–12 Science Education* recommends that by the end of grade 2, students understand that sound can make matter vibrate and that vibrating matter can make sound.

Your eardrum acts as a receiver of sound. When vibrating air reaches your eardrum, it causes your eardrum to vibrate as well. This in turn causes the tiny bones in the middle ear to vibrate, and when the vibrations reach the inner ear, they are turned into electrical signals and sent to the brain. When the vibrations are fast (a higher frequency), you hear a higher sound. When the vibrations are slow (a lower frequency), you hear a lower sound. Sound waves with a greater height (amplitude) will sound louder than sound waves with less amplitude. The human ear and brain working together are very good at recognizing and memorizing different sounds and decoding patterns in speech and music to distinguish them from random noise.

engage

What's That Sound? Read-Aloud and Sound Anticipation Guide

Inferring

Show students the cover of *What's That Sound?* by Mary Lawrence. *Ask*

? From looking at the cover, what do you think this book is about? (Students will likely infer that the book is about the sounds from a haunted house.)

Anticipation Guide

Tell them that before you read, you would like to find out what they already know about sound. Pass out the Sound Anticipation Guide and have students mark whether they think the statements are true or false in the "Before" column. Tell students that if they do not know the answer, just make their best guess. They will learn the correct answers in this lesson and will have a chance to answer them again in the "After" column.

> Connecting to the Common Core
> **Reading: Literature**
> Integration of Knowledge and Ideas: K.7, 1.7, 2.7
> Key Ideas and Details: K.1, 1.1, 2.1

Read aloud *What's That Sound?* skipping the insets as you read. Pause at the pages noted below to have students infer where the sounds are coming from. Encourage them to use clues from the text and illustrations to make their inferences.

After reading page 5, *ask*

? What do you think is making the high squeaky sound?

Then read the answer on page 6 (baby birds crying for food).

Read pages 7–9, then *ask*

? What do you think is making the soft tapping noise? Then read the answer on page 10 (a moth bumping against the screen).

Read pages 10–11, then *ask*

? Why do you think the window rattled?

Then read page 12, which explains that the window is shaking because sound waves from the thunder moved through the air to make the window vibrate. *Ask*

? What does the word vibrate mean? (to move back and forth)

Read pages 13–17, then *ask*

? What do you think caused the smashing sound from downstairs?

Then read the answer on pages 18–21 (Dad dropped a plate).

Read pages 22–27, then *ask*

? What do you think is making the low moaning noise?

Then read the answer on page 28 (Mr. Hubber playing the tuba).

Read pages 29–31, then *ask*

? Why do you think Mrs. Hubber had cotton in her ears? (Students should realize that the cotton muffled the sound of the tuba.)

After reading, reread page 10 and *ask*

? How did Amy know that the soft tapping sound was a moth? (She remembered the sound from camp last summer.)

Read the inset that explains how the ears and brain work together to recognize sounds we have heard several times. Tell students you would like to do a listening activity to see if their ears and brains can work together to recognize some sounds in the classroom.

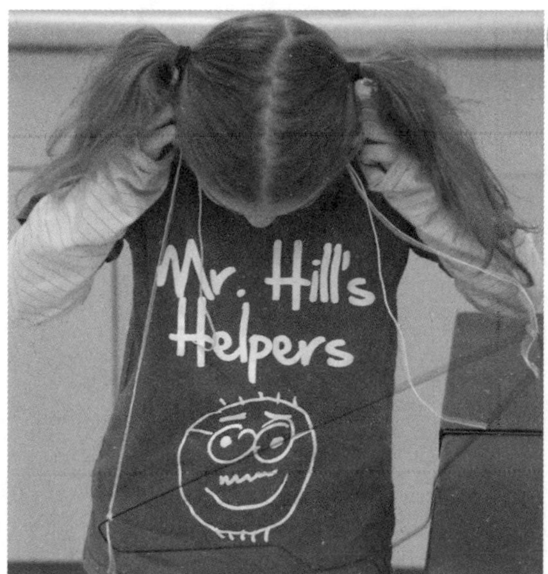

SOUND HANGER

explore

Listening Activity

Ask students to close their eyes and listen to the sounds around them. Circulate around the room and use classroom items to create sounds, such as using the pencil sharpener, crumpling up a piece of paper, bumping into the trash can, and so on. After one minute, ask them to open their eyes and describe what they heard:

? What did you hear?

? Where were the sounds coming from?

? What do you think made those sounds?

? How were you able to hear those sounds?

Sound Stations

In advance, set up seven sound stations around the room (see "Materials" section). Give each student a copy of the Sound Stations student page, and explain that they are going to be exploring sound at several different stations around the room. At each station, they can use words and pictures to answer the questions and record their observations. Provide time for students to experience each station in a small group.

explain

Sound Stations Discussion

Connecting to the Common Core
Speaking and Listening
COLLABORATION AND COMPREHENSION: K.1, 1.1, 2.1

When students have completed each station, bring them together with their student pages for a discussion. The following explanations are for you; modify as necessary to help students understand these concepts: sound can make things vibrate; sound can travel through solids, liquids, and gases; and the brain can recognize sounds.

Station 1: Feel the Beat

Sound can make things vibrate. The vibrations produced by beating the drum make the grains of rice move back and forth, or vibrate. The closer the drum, the more they vibrate. The farther the drum, the less they vibrate.

Station 2: Sound Hanger

Sound can travel through solids. The sound heard when placing the string-wrapped fingers in your ears is much louder than without it. The vibrations travel through the hanger and the string and make the hanger banging against the desk sound like a gong.

CARDBOARD KAZOOS

Station 3: Underwater

Sound can travel through liquids. When the rocks are tapped together in the air, the sound waves travel through the air and into your ear. But when the rocks are tapped together underwater, the sound waves travel through the water, then through the air and into your ear.

Station 4: Cardboard Kazoo

Sound can make things vibrate. Students should be able to feel the vibrations on the tissue paper at the end of the cardboard tube.

Station 5: Knock, Knock

Sound can travel through solids. The knock should be louder when you listen through the desk because the sound travels better through the solid desk than it does through the air.

Station 6: The Doctor Is In

Sound can travel through solids. Students should be able to hear their partner's heartbeat using the tube and funnel. The vibrations from the beating heart travel through the partner's chest wall, through the tube and funnel, and into the other partner's ear.

Station 7: Mystery Sounds

The brain can recognize sounds. Some sounds will be easier to guess if students have had prior experiences with the sounds.

Sounds All Around Read-Aloud

Connecting to the Common Core
Reading: Informational Text
KEY IDEAS AND DETAILS: K.1, 1.1, 2.1

Show students the cover of *Sounds All Around* and tell them that this nonfiction book can help them learn more about sound and help them fill out the "After" answers on their Sound Anticipation Guide.

PHONE A FRIEND

 *STOP and TRY IT*

After reading on page 8 "Feel your throat as you sing, talk, or hum," have students press gently on the middle of their throats and hum. *Ask*

? What do you feel? (Students should feel their fingertips tingle as their vocal chords vibrate.)

After reading page 9, have them be silent and feel their throats again. *Ask*

? Now, what do you feel? (nothing, their vocal chords are not vibrating)

After reading on page 19 "Sound waves travel through solid ground as well as air," have students place one ear on their desk and knock gently on the desk. *Ask*

? What do you hear? (a loud knock)

evaluate

Have students revisit the Sound Anticipation Guide from day 1. Ask them to think about the things they learned from the Sound Stations activity and the read-alouds in order to fill out the "After" column of the Sound Anticipation Guide.

Inquiry Place

Have students brainstorm questions about sound. Examples of such questions include

? Does the size of the cup or the type of string affect how well a paper cup telephone works? Test it!

? What materials muffle sound best? Test it!

? How do musical instruments make sound? Research it!

Then have students select a question to investigate or research as a class, or have groups of students vote on the question they want to investigate or research as a team. Have students present this information on a poster. Students will then share their findings in a poster session or gallery walk.

Discuss the answers together, having students cite examples and evidence from the hands-on activities and the books to support their answers. The correct answers to the anticipation guide are as follows:

1. Sound is made by vibrations.
   TRUE
2. You can see sound waves.
   FALSE
3. Sound can make things vibrate.
   TRUE
4. Sound waves can only travel through the air.
   FALSE
5. Sound is measured in decibels.
   TRUE
6. Some sounds can damage your ears.
   TRUE

elaborate and evaluate

Phone a Friend

Writing

> Connecting to the Common Core
> **Writing**
> Research to Build and Present Knowledge: K.8, 1.8, 2.8

Distribute the Phone a Friend student page and divide students into pairs. Have pairs of students follow the instructions on the student page to make a phone out of paper cups, paper clips, and string. *Note:* The string between the cups must be pulled tight for the phones to work properly.

Have students answer the questions on the student page and discuss their answers as a class:

1. Did the whispers travel better through the string or the air? (the string)
2. Can sound travel through a solid? (yes)
3. How do you know? (The sound traveled through the string and the string is solid.)
4. All sound is made by (vibrations, things vibrating, vibrating objects, or vibrating matter)

More Books to Read

Rosinsky, N. 2006. *Sound: Loud, soft, high, and low.* Mankato, MN: Picture Window Books.

Summary: This book describes how sound is created through vibrations that vary in pitch and volume.

Showers, Paul. 1993. *The Listening Walk.* New York: Harper Collins.

Summary: A father and child take a walk together and listen to the sounds around them.

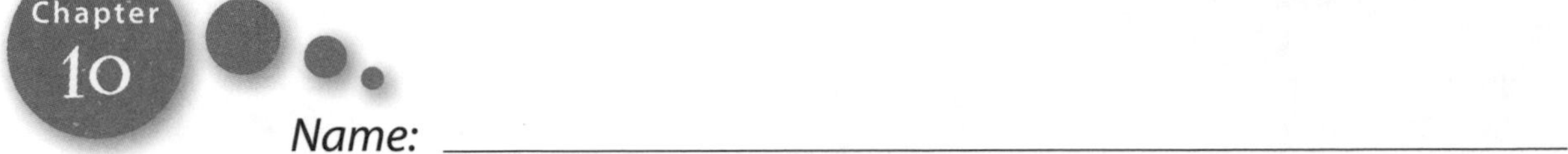

Name: ______________________________

Sound Anticipation Guide

Before True or False		After True or False
	1. Sound is made by vibrations.	
	2. You can see sound waves.	
	3. Sound can make things vibrate.	
	4. Sound waves can only travel through air.	
	5. Sound is measured in decibels.	
	6. Some sounds can damage your ears.	

Name: ____

Sound Stations

Station 1: Feel the Beat

Directions:

- Sprinkle some grains of rice on the plastic wrap covering the bowl.
- Hold the drum near, but not touching the plastic wrap, then hit the drum with the mallet.
- Observe the rice grains.
- What happens when you hit the drum harder? ______________

____

- What happens when you put the drum farther away and hit it?

____

____

____

____

Station 2: Sound Hanger

Directions:

- Wrap the ends of the strings tied to the hanger around your fingers.
- Bang the hanger against the desk.
- Then place your fingers in your ears and bang the hanger gently against the desk.
- How did the sound change when you put your fingers in your ears?

Station 3: Underwater

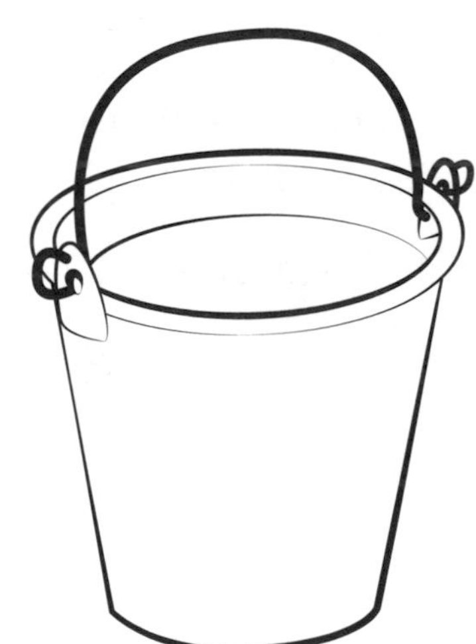

Directions:

- Tap the rocks together and listen to the sound they make.
- Hold the rocks underwater and tap them together again.
- How are the sounds different?

Station 4: Cardboard Kazoo

Directions:

- Write your name on a paper tube.
- Put a piece of tissue paper over one end of the tube and attach it with a rubber band.
- You have made a kazoo!
- Make different sounds by humming while you hold the open end of the kazoo up to your mouth.
- Gently put your fingers on the tissue paper while you are humming.
- What does it feel like?

____

____

Station 5: Knock, Knock

Directions:

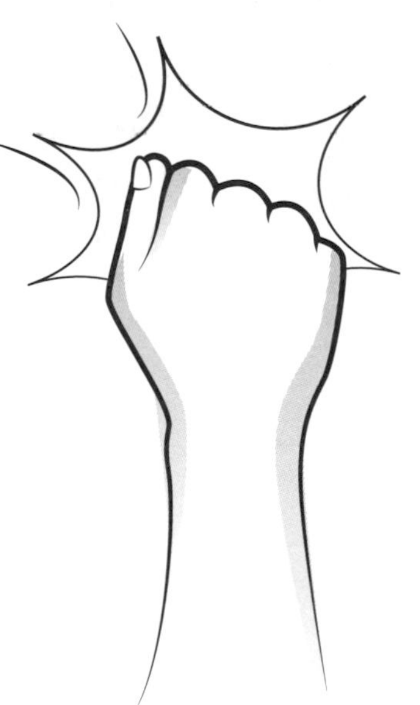

- Have your partner sit on the other side of the desk and gently knock on it.
- Then, place your ear on the desk as your partner gently knocks again.
- Try it several times.
- Which is louder?

____

____

Station 6: The Doctor Is In

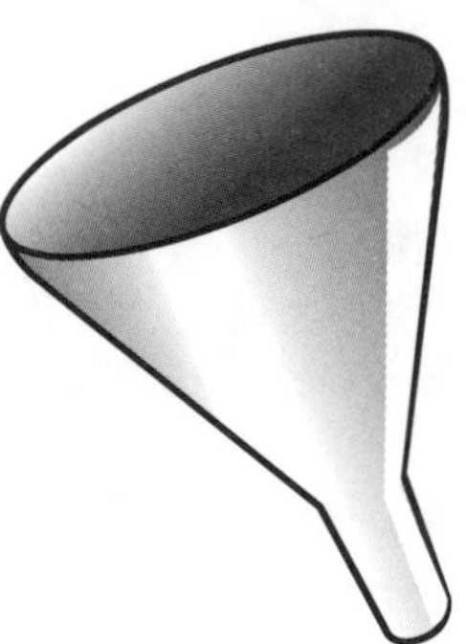

Directions:

- Hold the funnel on your ear and have your partner hold the end of the plastic tubing to his or her chest.
- What do you hear?

____

____

Station 7: Mystery Sounds

Directions:

- Shake each mystery container gently and try to guess what is inside.
- Record your guesses in the table on the next page.
- After you have all had a chance to guess, carefully open each mystery container and look inside. Record what was in the container in the table on the next page.

Station 7: Mystery Sounds *continued*

Container	My Guess	What Was Inside
A		
B		
C		
D		
E		
F		

- Which were easier to guess?

____

____

____

- Which were harder to guess?

____

____

____

Name : ______________________________

Phone a Friend

Can sound travel through a solid, like string? Let's find out!

Materials:

3 meters of cotton string

2 plastic cups with holes in the bottom

2 paper clips

Directions:

- Place the ends of the string through the holes in each cup.
- Tie a paper clip to the string inside the cup to keep the string from being pulled out of the hole.
- Pull the string tight between the cups.
- Hold a cup to your ear and ask your partner to whisper into the other cup.
- Take turns talking through the "telephone."
- Next, take turns whispering to each other without using the "telephone."

Phone a Friend *continued*

1. Did the whispers travel better through the string or the air?

____

____

____

2. Can sound travel through a solid?

____

____

____

3. How do you know?

____

____

____

____

4. All sound is made by

____

Do You Know Which Ones Will Grow?

Description

How can you tell if something is alive? What do living things need? These questions and many more are answered through activities and picture books in this lesson for young children.

Suggested Grade Levels: K–2

LESSON OBJECTIVES *Connecting to the Framework*

LIFE SCIENCES

Core Idea LS1: From Molecules to Organisms: Structures and Processes

LS1.C: Organization for Matter and Energy Flow in Organisms

By the end of grade 2: All animals need food in order to live and grow. They obtain their food from plants or from other animals. Plants need water and light to live and grow.

Core Idea LS4: Biological Evolution: Unity and Diversity

LS4.D: Biodiversity and Humans

By the end of grade 2: There are many different kinds of living things in any area, and they exist in different places on land and in water.

Featured Picture Books 

TITLE: ***Do You Know Which Ones Will Grow?***
AUTHOR: **Susan A. Shea**
ILLUSTRATOR: **Tom Slaughter**
PUBLISHER: **Blue Apple Books**
YEAR: **2011**
GENRE: **Non-Narrative Information**
SUMMARY: *Rhyming text and fun lift-the-flap illustrations ask the reader which things will grow, such as "If a duckling grows and becomes a duck, can a car grow and become a … truck?"*

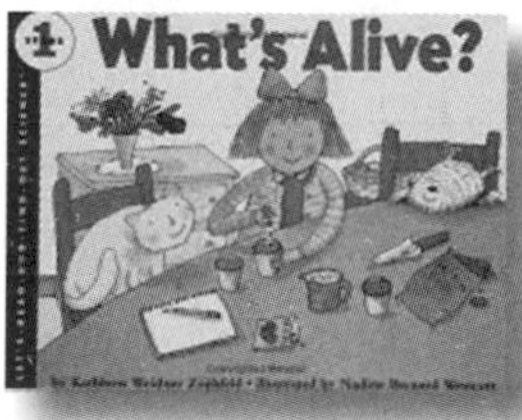

TITLE: ***What's Alive?***
AUTHOR: **Kathleen Weidner Zoehfeld**
ILLUSTRATOR: **Nadine Bernard Westcott**
PUBLISHER: **HarperCollins**
YEAR: **1995**
GENRE: **Narrative Information**
SUMMARY: *From the* Let's-Read-and-Find-Out Science *series, this book introduces students to the differences between living and nonliving things.*

Time Needed

This lesson will take several class periods. Suggested scheduling is as follows:

Day 1: **Engage** with Open Sort, *Do You Know Which Ones Will Grow?* Read-Aloud, and Closed Sort, and **Explore** with Things That Grow Indoors and Outdoors

Day 2: **Explain** with *What's Alive?* Read-Aloud

Day 3: **Elaborate** with Looking for Living and Nonliving Things

Day 4: **Evaluate** with Is It Alive? Lift-the-Flap Booklet

Materials

For Open Sort and Closed Sort (per group of three to four students)

- Picture cards

For Things That Grow Indoors and Outdoors (per student)

- Clipboard

> **NOTE:** The explore and elaborate phases of this lesson require an outdoor area.

For Looking for Living and Nonliving Things

Per student

- Hand lens
- Clipboard

Per group of three to four students

- Hula hoops or string to make a circle
- Selection of nonliving things (e.g., balls, toys, and rocks) to place in the hula hoops or string circles

For Is It Alive? Lift-the-Flap Booklet (per student)

- Magazines
- Scissors
- Markers or crayons
- Glue

Student Pages

- Things That Grow (booklet; copy p. 154 back-to-back with p. 155)
- Looking for Living and Nonliving Things
- Is It Alive? (lift-the-flap booklet; copy p. 157 back-to-back with p. 158)

Background

The difference between living and nonliving things is an essential concept for elementary students to understand. It may seem like a simple concept, but it can be tricky for young children. Scientists have developed a set of criteria for determining whether or not something can be considered living. Living things grow, change, reproduce, and have certain needs. However, many nonliving things that children encounter might appear to have one or more of these qualities. *A Framework for K–12 Science Education* suggests that students learn at an early age that living things have needs that must be met for them to survive. Animals need air, water, and food. Plants need air, water, light, and nutrients, but they do not need to eat food. Instead they make their own food. Plant nutrients are chemicals that are important to a plant's growth, but nutrients are not the same as food. They are dissolved in water and absorbed through the plant's roots. Many students have the misconception that fertilizer is the same as food. They may have even seen fertilizer at the store or at home that is labeled "plant food." It is important for students to understand that fertilizer contains nutrients, but not food.

The main difference between plants' and animals' needs is that plants do not eat other living things to get energy. They make their own food from air and water with energy from sunlight, in a process called photosynthesis. The process of *photosynthesis* might be too complex for very young children to understand. If students learn that plants do not eat any living things to get energy, that is sufficient to build the foundation for later understanding of the complex process that plants use to make their own food.

engage

Do You Know Which Ones Will Grow?

Open Sort

Give each pair of students a set of picture cards and ask them to look at each picture and come up with a way of sorting them into groups. When they have finished sorting, have pairs guess how other pairs nearby sorted the pictures by looking at their groups. Discuss the different ways of sorting the pictures and explain that at this point there was no right or wrong way to sort.

Do You Know Which Ones Will Grow? Read-Aloud

Inferring

> Connecting to the Common Core
> **Reading: Literature**
> Key Ideas and Details: K.1, 1.1, 2.1

Show students the cover of *Do You Know Which Ones Will Grow?* and introduce the author and illustrator. Tell students to listen and watch for the items from their cards as they appear in the book. Read aloud the first few pages to give students a feel for the pattern and rhyme, then ask students to infer from the illustrations and the text what is under each flap. For example, after reading, "If a kit grows and becomes a fox, can a watch grow and become ..." ask students to predict what is under the flap. Then open the flap to reveal a clock.

Closed Sort

After reading, ask students to sort their cards into two groups: Grows and Does Not Grow. *Ask*

? What kinds of things grow? (things that are alive)

? What else do living things do besides grow?

? How can you tell if something is alive?

explore

Things That Grow Indoors and Outdoors

Tell students that you are going to look for more things that grow. Give each student a copy of the Things That Grow booklet and a clipboard. Tell them that on the left-hand side of the page they are going to make a list of things in the classroom that grow. They can record this list in words or pictures. Allow them time to walk around the room quietly with their clipboards, booklets, and pencils and look for things that grow. If your classroom does not contain any plants or animals, they likely will not have much to list—except themselves! Bring students back together and have them share some of the items on their lists.

Next, tell students that you are going to do the same exercise, but this time they will be looking for things that grow outdoors and will record their list on the right-hand page. Allow them time to walk around an area outdoors with their clipboards, booklets, and pencils and look for things that grow. Then bring students back together and have them share some of the items on their lists.

SAFETY

- Students should wear closed-toe shoes or sneakers, long pants, long-sleeve shirts, hats, sunglasses, sunscreen, and safety glasses or goggles when working outdoors.
- Caution students to watch out for ticks, mosquitoes, stinging insects, and other potentially hazardous insects when working outdoors.
- Caution students against poisonous plants such as poison ivy or poison sumac when working out-of-doors.
- Check with the school nurse regarding student medical issues (e.g., allergies to bee stings) and how to deal with them.
- Find out whether outdoor areas have been treated with pesticides, fungicides, or any other toxins, and avoid any such areas.
- Bring some form of communication, such as a cell phone or two-way radio, in case of emergencies.
- Inform parents in writing of a planned field trip, any potential hazards, and the safety precautions being taken.
- Have students wash their hands with soap and water upon completing the activity (before and after when consuming food).

Ask

? Were there more things that grow indoors or outdoors? (outdoors) Why?

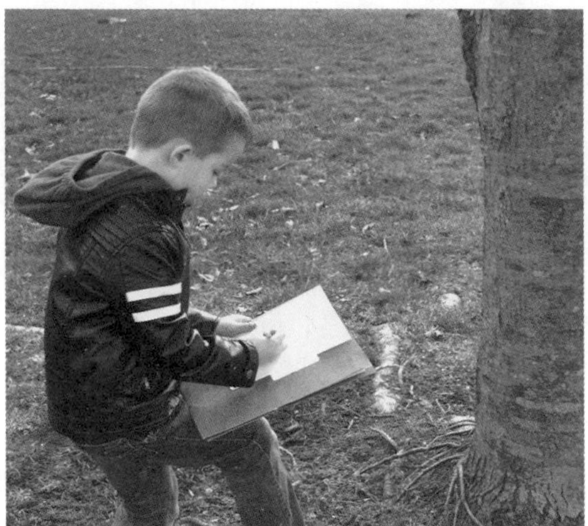

LOOKING FOR THINGS THAT GROW OUTDOORS

? What do these things need to grow?

? What else do these things have in common? (They are alive.)

? How can you tell if something is alive?

explain

What's Alive? Read-Aloud

Connecting to the Common Core
Reading: Informational Text
KEY IDEAS AND DETAILS: K.1, 1.1, 2.1

Determining Importance

Tell students that you have a book that will help them learn more about how to tell if something is alive. Show students the cover of *What's Alive?* by Kathleen Weidner Zoehfeld. Tell students that as you read the book aloud, you would like them to listen for all of the ways that living things are alike. Read the book aloud, pausing periodically to point out the ways in which living things are alike:

- Living things need food.
- Living things need water.

- Living things need air.
- Living things grow.
- Living things move.

After reading page 16, which says that plants "need water, air, and food. And they can move and grow," *ask*

- ? Do plants eat food the way animals do? (no)
- ? Then how do they get food?
- ? Do they move around the way animals do? (No, they do not run, jump, or fly.)
- ? Then how do they move?

Let's read on to find out.

After reading page 20, discuss how plants get food (they make it) and how they move (grow and bend).

Explain to students another trait that all living things have is the ability to reproduce, or make more of themselves. For example, the cat on pages 12 and 13 had kittens, the bird on pages 14 and 15 hatched from an egg its mother laid, and the trees and flowers on page 18 made seeds that will grow into new flowers and plants.

Then read the rest of the book aloud.

Questioning

Ask

- ? What do we call things that are not living things? (nonliving things)
- ? What were some examples of nonliving things from the book? (stone, tricycle, book, doll, etc.)
- ? What if you find a brown, dried-up plant or an insect that is not moving anymore? Is it living or nonliving? (They are living things that have died.)
- ? What do plants and animals have in common? (They are living; they grow; they change; they reproduce; they need air, food, and water.)
- ? How are plants and animals different? (Plants are often green; they don't move around like animals; many plants make seeds instead of laying eggs or having babies.)

Tell students that plants and animals are also very different in the way that they get food. Explain that animals eat plants or other animals to get energy, but plants make their food out of water and air with energy from the Sun. So plants do not "eat" anything. A common misconception with young students is that plants take in food from the soil. This is incorrect. Explain to students that plants can get nutrients from the soil or fertilizer, but this is not food. It is similar to how people take vitamins. They give us nutrients, but we could not survive on them because they are not food.

Determining Importance

> Connecting to the Common Core
> **Reading: Informational Text**
> Key Ideas and Details: K.2, 1.2, 2.2

Ask

- ? If a friend asked you what this book is about, what would you tell them? In other words, what is the main topic of the book? (Have students turn and talk. Students should recognize that the main topic of the book is how to tell what's alive and what's not.)

elaborate

Looking for Living and Nonliving Things

Ahead of time, set up a hula hoop or circle of string for each group of three or four students in a grassy area outdoors. In each hula hoop or circle be sure there are examples of several living things (grass, insects, flowers, and so on) and nonliving things (ball, toy, rock, and so on). Give each student a copy of the Looking for Living and Nonliving Things student page, a clipboard, and a hand lens. Divide students into groups and assign each group a hooped area (circle) to quietly explore. Have students draw and label

Looking for living and nonliving things

what they see in their circle and make a list of all of the living things they see and all the nonliving things in the circle. Visit groups as they are working and ask them to point out some living and nonliving things to you. Encourage them to use their hand lenses to look very closely. Ask guiding questions, such as

- ? How do you know that is living/nonliving?
- ? Does it need air, food, and water?
- ? Does it grow?
- ? Does it move?

Students may find some dried twigs, brown leaves, or dead insects in their circles. Explain to them that these items do not fit into either category on the student page. They are in another category called once-living things.

After returning to the classroom, have students share their drawings and their living and nonliving lists.

evaluate

Is It Alive? Lift-the-Flap Booklet

Writing

Connecting to the Common Core
Writing
Research to Build Knowledge: K.8, 1.8, 2.8

Tell students that they are going to help you create an "Is It Alive?" bulletin board. Revisit the book *Do You Know Which Ones Will Grow?* and remind students how the flaps in the book worked. Give each student the Is It Alive? lift-the-flap student page folded on the dotted line, some magazines, markers or crayons, glue, and scissors.

SAFETY

Use caution in working with sharp items like scissors, because they can cut or puncture skin.

Inquiry Place

Have students brainstorm questions about living things. Examples of such questions include

- ? How are living things classified, or sorted, by scientists? Research it!
- ? What other living things are there besides plants and animals? Research it!
- ? Does a seed need sunlight to sprout? Test it!
- ? Can you get a plant to bend toward a light? Try it!

Then have students select a question to investigate or research as a class, or have groups of students vote on the question they want to investigate or research as a team. Have students present this information on a poster. Students will then share their findings in a poster session or gallery walk.

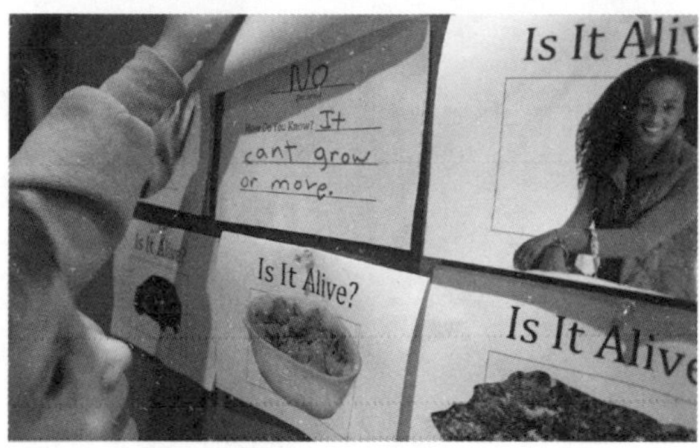

Lift-the-Flap Booklet

Tell students that they are going to be creating something similar to the flaps in the book. On the outside of their booklet, they will glue or draw a picture of something living or nonliving. On the inside of the booklet they will write "yes" if the thing is living and "no" if it is not. They will also answer the question, "How do you know?" Possible answers are:

Nonliving

No. It does not need food, air, or water.

No. It does not move or grow.

No. It does not make more of itself (reproduce).

No. It cannot die.

Living

It grows and moves.

It needs air, food, and water.

It can make more of itself (reproduce).

It can die.

Evaluate student understanding by checking that they correctly identified the picture as living or nonliving and assessing their answer to the question, "How do you know?" inside the booklet. Display all of the booklets on a bulletin board with the title "Is It Alive?" Invite students to look at their classmates' booklets and decide if each thing is alive or not and check their answers under the flaps.

More Books to Read

Kalman, B. 2008. *Is it a living thing?* New York: Crabtree.
Summary: This book details the characteristics of living things, including that they are made of cells and have life cycles. Includes captions and an index.

Lindeen, C. 2008. *Living and nonliving.* Mankato, MN: Capstone Press.
Summary: Simple text and photographs introduce young students to the differences between living and nonliving things.

Rissman, R. 2009. *Is it living or nonliving?* Chicago: Heinemann Library.
Summary: This Acorn Read-Aloud title explains how to tell if something is living. Includes an index, bold-print words, and glossary.

Royston, A. 2008. *Living and nonliving.* Chicago: Heinemann Library.
Summary: This book from the *My World of Science* series describes the difference between living and

nonliving things. Includes an index, bold-print words, and glossary.

Silver, D. M. 1997. *One small square: Backyard.* New York: McGraw-Hill.
Summary: This book teaches children that a small square of earth can yield up an endlessly complex and fascinating interaction of plants and animals with their environment, and shows them how to study the area as a scientist would.

Picture Cards

Picture Cards

Picture Cards

Picture Cards

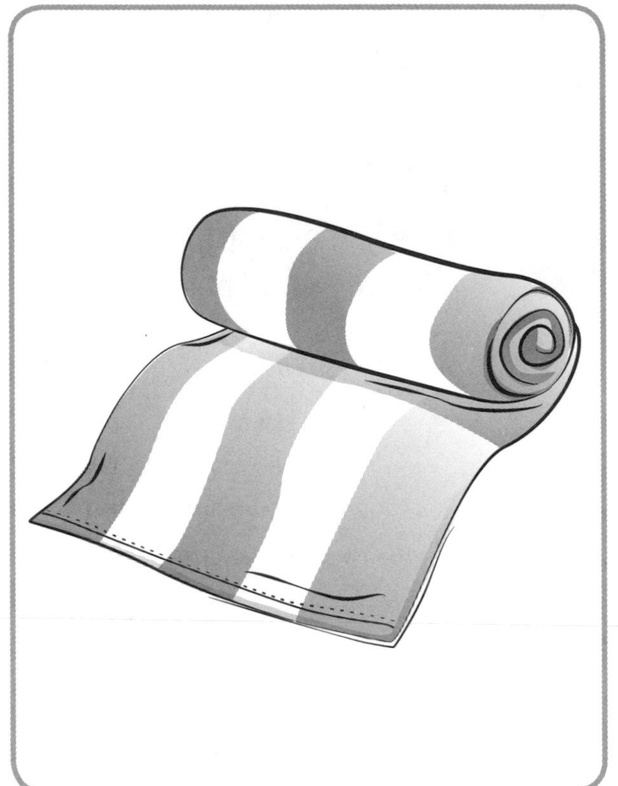

Picture Cards

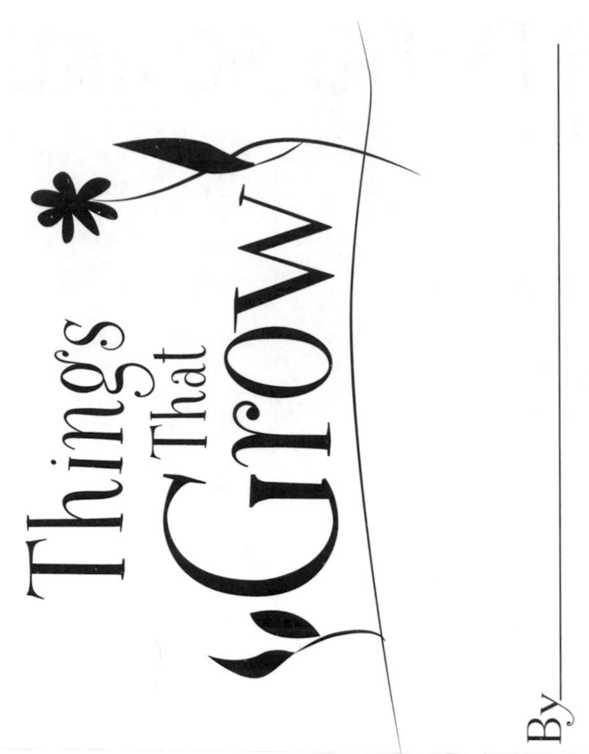
Things
That
Grow
By

Things That Grow

In the Classroom

Things That Grow

Outdoors

Name: ____________________

Looking *for* Living and Nonliving Things

Draw what you see in your circle.

Living

____

____

Nonliving

____

____

(yes or no)

How Do You Know?

Is It Alive?

Seeds on the Move

Description

This lesson explores the variety of ways that plants disperse their seeds to produce more plants. Learners read about how wind and water help spread seeds as well as many fascinating ways animals "plant"trees. Then students take a "sock walk" to explore how plants sometimes depend on animals to move their seeds around.

Suggested Grade Levels: K–2

LESSON OBJECTIVES *Connecting to the Framework*

LIFE SCIENCES

Core Idea LS1: From Molecules to Organisms: Structures and Processes

LS1.A: Structure and Function

By the end of grade 2: All organisms have external parts. ... Plants ... have different parts (roots, stems, leaves, flowers, fruits) that help them survive, grow, and produce more plants.

Core Idea LS2: Ecosystems: Interactions, Energy, and Dynamics

LS2.A: Interdependent Relationships in Ecosystems

By the end of grade 2: Animals can move around, but plants cannot, and they often depend on animals for pollination or to move their seeds around.

Featured Picture Books

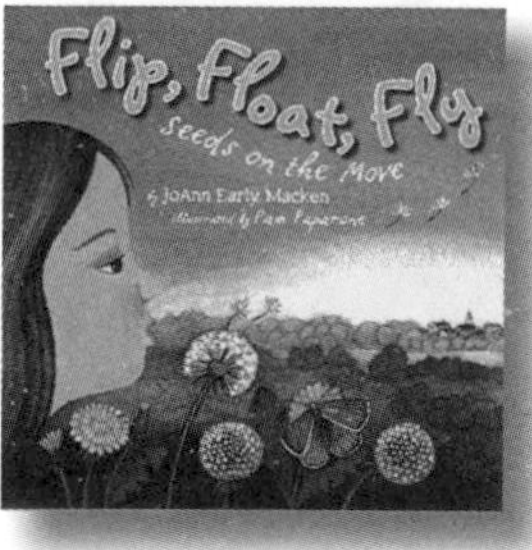

TITLE: ***Flip, Float, Fly: Seeds on the Move***
AUTHOR: **JoAnn Early Macken**
ILLUSTRATOR: **Pam Paparone**
PUBLISHER: **Holiday House**
YEAR: **2008**
GENRE: **Non-Narrative Information**
SUMMARY: *Colorful paintings and simple text explain many different methods of seed dispersal. The end matter includes information on plant parts and explains that not all seeds sprout.*

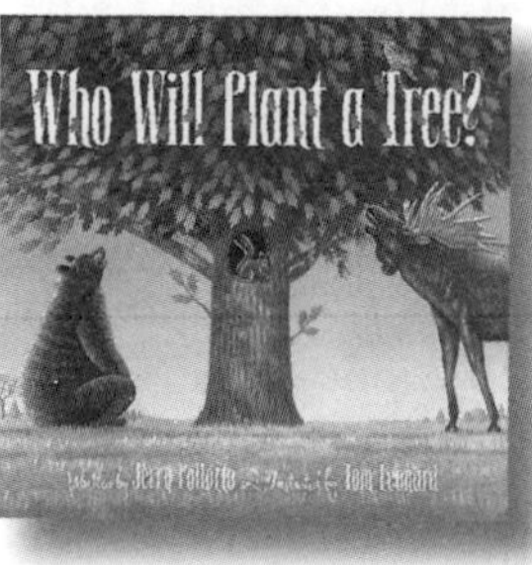

TITLE: ***Who Will Plant a Tree?***
AUTHOR: **Jerry Pallotta**
ILLUSTRATOR: **Tom Leonard**
PUBLISHER: **Sleeping Bear Press**
YEAR: **2010**
GENRE: **Narrative Information**
SUMMARY: *This funny, informative book depicts the ways seeds are dispersed by various animals' behaviors and body structures.*

Time Needed

This lesson will take several class periods. Suggested scheduling is as follows:

Day 1: **Engage** with a discussion of dandelions, **Explore** with Seed Observations, and **Explain** with *Flip, Float, Fly* Read-Aloud

Day 2: **Elaborate** with *Who Will Plant a Tree?* Read-Aloud and Sock Walk

Day 3: **Evaluate** with Seeds on the Move Lift-the-Flap Booklet

Materials

For Seed Observations (per group of four students):

- Collection of seeds featured in the book *Flip, Float, Fly,* such as dandelion, maple, locust, coconut, acorn, burr, or whole fig cut in half to reveal the seeds (Optional: Use seed cards.)
- Hand lenses

For Sock Walk (per student):

- Large white or light-color sock (athletic or fluffy socks work best)
- Hand lens
- (Optional) Gallon-size zippered plastic bag

NOTE: The elaborate phase of this lesson requires an outdoor area, preferably containing tall, unmown grass or weeds.

Student Pages

- (Optional) Seed Cards
- What's on My Sock?
- Seeds on the Move (lift-the-flap booklet)

Background

A Framework for K–12 Science Education suggests that by the end of grade 2 students should understand that plants have different parts that help them survive, grow, and produce more plants. One of those parts is a seed. Inside each seed is an embryo of a plant that, if conditions are right, will grow into a new plant. Multiple seeds often cannot survive very near the parent plant because too many seedlings would crowd the same spot, layers of leaves from the parent plant might block the light, or too many roots would compete for water. Thus, it is key for the survival of many plant species that the seeds be dispersed to other places.

There are many ways that seeds are moved from place to place. Sometimes it is as simple as the wind blowing them off the plant and carrying them far away, like the dandelion seed or the maple seed. These two seeds have parts that allow them to be easily carried by the wind. The dandelion has a feathery tip that catches in the wind and floats easily. The maple seed has a winglike structure that allows it to sail on the wind and twist and twirl through the air. Sometimes water moves seeds from place to place. For example, floating coconuts can be carried by ocean currents to other pieces of land where they can sprout and grow into new coconut trees. Some plants depend on animals for pollination or to move their seeds around. Many species of animals carry seeds from place to place without even knowing it. Seeds can stick to their fur or feathers and fall off in another location. Some animals eat fruits and the seeds move through their digestive systems, ending up in their droppings. In these ways, new plants can grow far away from the tree or plant that produced the fruit.

engage

Making Connections: Text-to-Self

Show students the cover of *Flip, Float, Fly,* which shows a girl blowing the seeds off a dandelion. *Ask*

? Have you ever seen these fuzzy things growing? What are they? (dandelion seed heads)

? Have you ever made a wish and blown on them? What happened to the seeds? (They floated away.) Did you know you were helping spread dandelion seeds so that more dandelions could grow?

Explain that plants have many ways of spreading their seeds so that more plants can grow. Tell students that they are going to observe a variety of seeds and then try to figure out how they move from place to place.

OBSERVING SEEDS

SAFETY

- When working with seeds, use only sources that are pesticide/herbicide free.
- Check with the school nurse regarding student medical issues (e.g., allergies to pollen or tree nuts) and how to deal with them.

explore

Seed Observations

In advance, collect some of the seeds featured in the book *Flip, Float, Fly* for each group of four students to observe. Examples of these seeds are dandelion, maple, locust, coconut, acorn, burr, and whole fig (the fig should be cut in half to reveal the seeds). Give each group of students a collection of the seeds to observe. Have them use hand lenses to make careful observations of their size, shape, and other characteristics. Having students observe real seeds is preferred, but if you are not able to collect the seeds, you can use the Seed Cards.

After the students have observed the seeds, *ask*

? How do you think each of these seeds might be spread to other places? (Have students turn and talk about the characteristics of each seed and share ideas of how these seeds might be moved from place to place.)

explain

Flip, Float, Fly Read-Aloud

Connecting to the Common Core
Reading Informational Text
Key Ideas and Details: K.1, 1.1, 2.1

Determining Importance

Tell students that you are going to read the book *Flip, Float, Fly* to find out how each of the seeds they observed travels from its parent plant. Have students signal when they hear about each of the seeds they observed as you read. Then read the book aloud, stopping before the "Notes" page at the end.

After reading, ask students to use evidence from the book to explain how the seeds they observed earlier move from place to place.

elaborate

Who Will Plant a Tree? Read-Aloud

Connecting to the Common Core
Reading: Informational Text
Key Ideas and Details: K.1, 1.1, 2.1

Inferring

Show students the cover of the book *Who Will Plant a Tree?* and *ask*

? What do you think this story is about? Why?

Encourage students to explain their thinking by referring back to the cover's illustration and title.

Point out the bear, squirrel, and moose on the cover. *Ask*

? Do you think these animals can help to plant a tree? How?

Determining Importance

Ask students to listen for the different ways that animals can help seeds travel from their parent plant to grow in other places as you read the book aloud.

Questioning

Read the book aloud. After reading, *ask*

? What were some of the ways the animals in the book planted trees? (Seeds stuck to their fur or feathers and then fell off later in a different place, they ate seeds and then pooped or spit them out, and so on.)

? Did these animals know they were planting trees? Did they do it on purpose? (no, except for the people at the end of the book)

Synthesizing

Connecting to the Common Core
Reading: Informational Text
Craft and Structure: K.6, 1.6, 2.6

Show students the back flap of the book and tell them that often you can find more information about the author and illustrator here. Read aloud the paragraphs about the author, Jerry Pallotta, and the illustrator, Tom Leonard. Then flip through the illustrations and reread some of the text from a few pages. *Ask*

? What do you think the author and illustrator of this book are trying to tell you about trees with their words and pictures? (that trees often depend on animals to move their seeds around)

Making Connections: Text-to-Text

Connecting to the Common Core
Reading: Informational Text
Integration of Knowledge and Ideas: K.9, 1.9, 2.9

Ask

? How does this book compare with the other book we read, *Flip, Float, Fly*? (*Flip, Float, Fly* was about wind, water, and animals moving seeds around. *Who Will Plant a Tree?* was about animals moving seeds around.)

Sock Walk

Ahead of time, ask each student to bring in a sock, preferably an adult-size sock (the fuzzier the better) that will easily fit over one of their shoes. Be sure to have some extra socks for any students who forget to bring them.

Ask

? Do you think animals around our school help to plant trees or other plants without knowing it?

? What kinds of seeds might they collect on their fur?

Tell students that they will be taking a walk outdoors to see if they can collect any seeds. But they won't be using their hands to collect the seeds! Instead, they will be placing a sock over one of their shoes to make a model of an animal's fur-covered leg. As they walk around the school grounds, different kinds of seeds might stick to the socks, just like the seeds in the book stuck to the animals. After the sock walk, they will be examining their socks to see if they collected any seeds.

The best location for a sock walk would be a large unmown area of grass or weeds. Be sure to check for poisonous plants ahead of time. After the walk, have students carefully remove their socks before moving indoors to examine what was collected. Once inside, ask students to use hand lenses to observe what was collected on their socks. Have students draw and label what they observed on the What's on My Sock? student page.

An alternate activity is to sprinkle some seeds on desktops in advance. Then have each student put a sock on one hand and do the sock walk indoors.

After the sock walk, *ask*

? Which things on your sock do you think are seeds? Why do you think so?

? How could you know for sure? (plant the seeds)

You may want to see if the seeds will sprout by wetting the seed-covered socks with water, plac-

SAFETY

- Students should wear long socks, long pants, long-sleeve shirts, hats, sunglasses, sunscreen, and safety glasses or goggles.
- Caution students against collecting ticks, mosquitoes, stinging insects, and other potentially hazardous insects.
- Check with the school nurse regarding student medical issues (e.g., allergies to bee stings) and how to deal with them.
- Find out whether outdoor areas have been treated with pesticides, fungicides, or any other toxins and avoid any such areas.
- Caution students agains poisonous plants such as poison ivy or poison sumac.
- Bring some form of communication, such as a cell phone or two-way radio, in case of emergencies.
- Inform parents, in writing, of planning field trip, any potential hazards, and having safety precautions being taken.
- Have students was their hands with soap and water upon completing the activity.

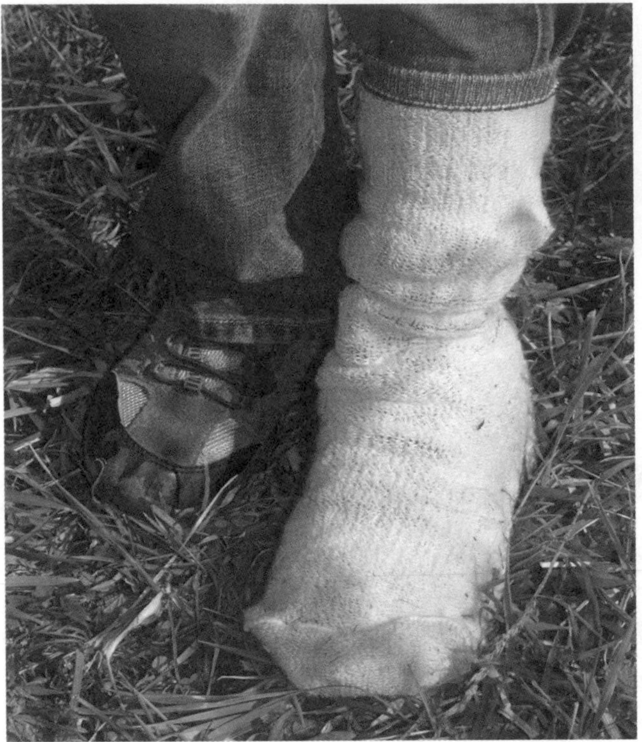

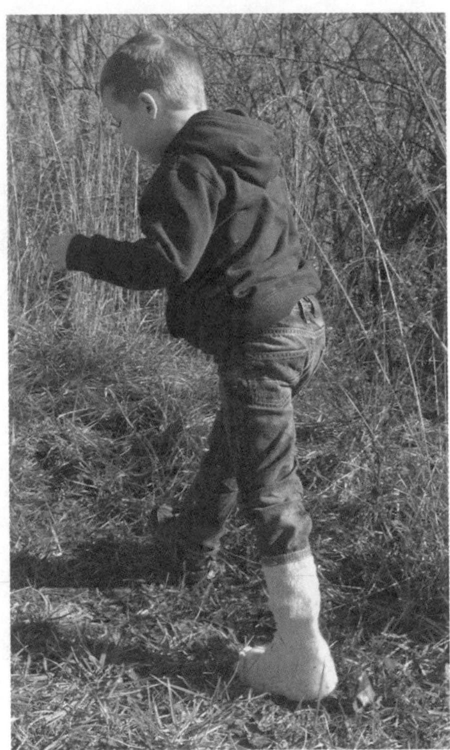

Taking a Sock Walk

ing each in a zippered bag, and keeping them in a warm place for a few weeks.

evaluate

Seeds on the Move Lift-the-Flap Booklet

Writing

> Connecting to the Common Core
> **Writing**
> Research to Build Knowledge: K.8, 1.8, 2.8

Tell students that they are going to have an opportunity to show what they have learned about how seeds are moved. Give each student a copy of the Seeds on the Move student page. To make a lift-the-flap booklet, have them fold each page on the dotted line and staple the pages. For each page, students should write the name of a seed that travels in that way and then draw the wind, water, or animal helping it move.

More Books to Read

Bodach, V. K. 2007. *Seeds.* Mankato, MN: Capstone Press.
Summary: Simple text and bold, close-up photographs present the seeds of different plants, how they grow, and their uses.

Gibbons, G. 1993. *From seed to plant.* New York: Holiday House.
Summary: This book provides a simple introduction to how plants reproduce. Topics include pollination, seed dispersal, and growth.

Jordan, H. J. 1992. *How a seed grows.* New York: HarperCollins.
Summary: This *Let's-Read-and-Find-Out Science* book provides a simple introduction to how seeds grow into plants.

Inquiry Place

Have students brainstorm testable questions about seeds. Examples of such questions include

- ? Will seeds sprout in the dark? Test it!
- ? Will seeds sprout in the freezer? Test it!
- ? Will seeds sprout without water? Test it!

Then have students select a question to investigate as a class, or have groups of students vote on the question they want to investigate as a team. After they make their predictions, have them design an experiment to test their predictions. Students can present their findings in a poster session or gallery walk.

Robbins, K. 2005. *Seeds*. Atheneum Books for Young Readers.
Summary: This book has stunning photographs and straightforward text that explains how seeds grow and how they vary in size, shape, and dispersal patterns.

Weakland, M. 2011. *Seeds go, seeds grow*. Mankato, MN: Capstone Press.
Summary: Simple text and photographs explain the basics of seed parts, how they are produced, and how they can be moved to different places by wind, water, and animals.

Seed Cards

Dandelion Seeds

Maple Seeds

Tumbleweed Seeds

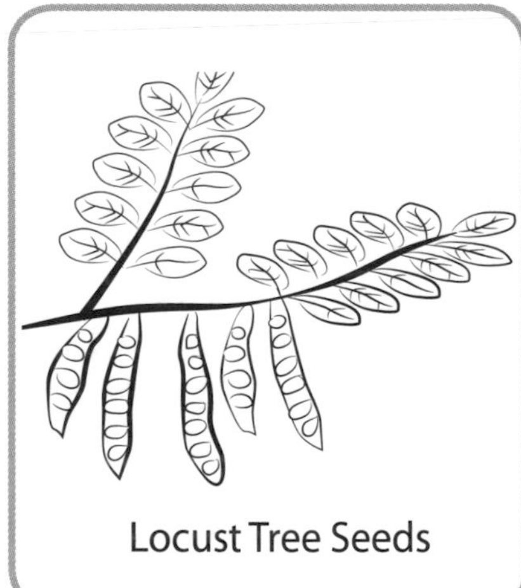
Locust Tree Seeds

Coconut Seeds

Wild Oat Seeds

Fig Seeds

Burdock seeds

Touch-Me-Not Seeds

Name ______________________________

What's on My Sock?

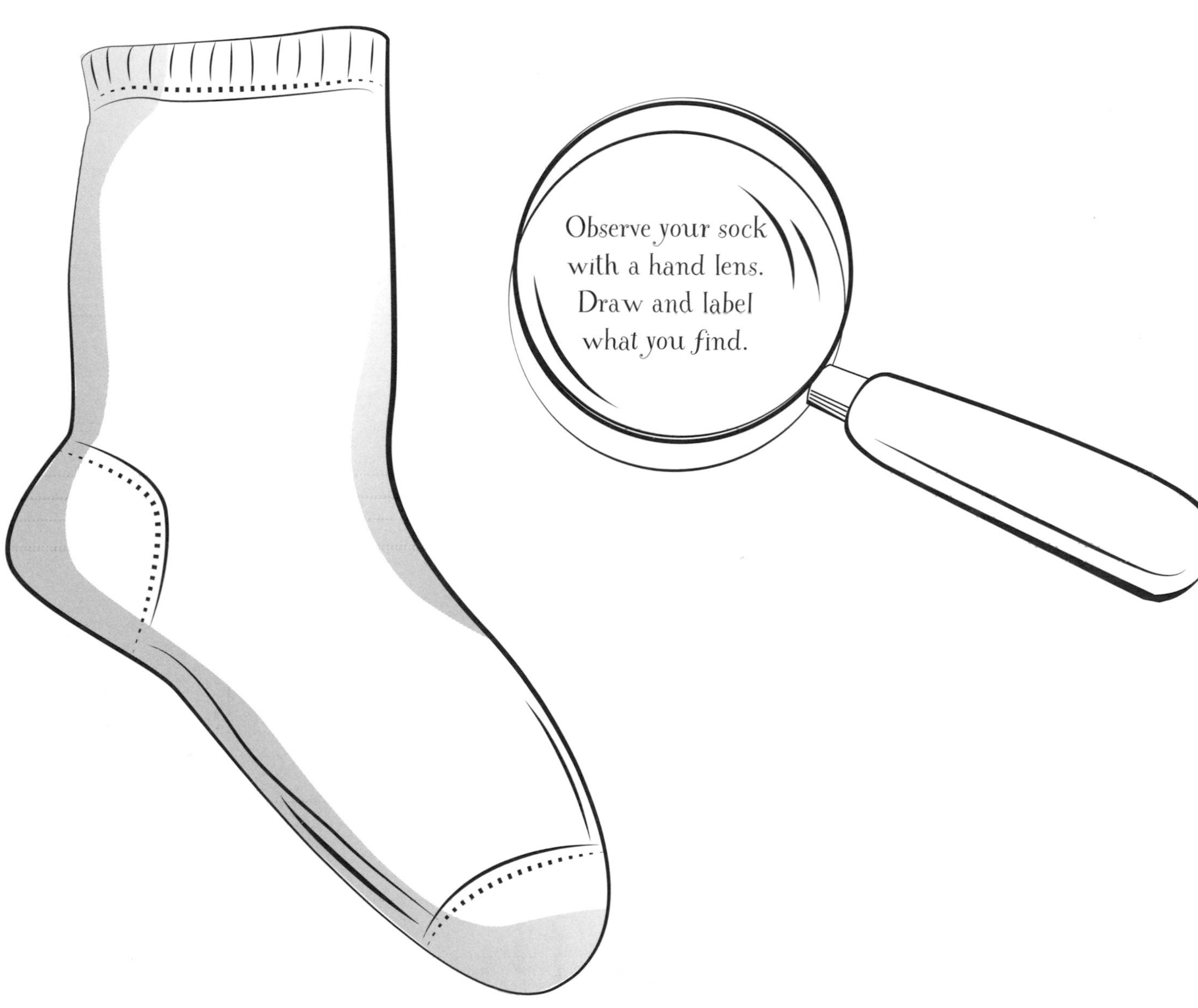

Seeds on the Move

By____________________________________

Water

Wind

Animals

Unbeatable Beaks

Description

Watching birds visit feeders outside your classroom can provide joy and delight for both you and your students. It can also be a valuable learning experience about the structure and function of living things. In this lesson, students observe local birds at a classroom feeding station and then explore, through a simulation activity, how different bird beaks are suited for different food sources.

Suggested Grade Levels: K–2

LESSON OBJECTIVES *Connecting to the Framework*

LIFE SCIENCES

Core Idea LS1: From Molecules to Organisms: Structures and Processes

LS1.A: Structure and Function

By the end of grade 2: All organisms have external parts. Different animals use their body parts in different ways to see, hear, grasp objects, protect themselves, move from place to place, and seek, find, and take in food, water, and air.

Featured Picture Books

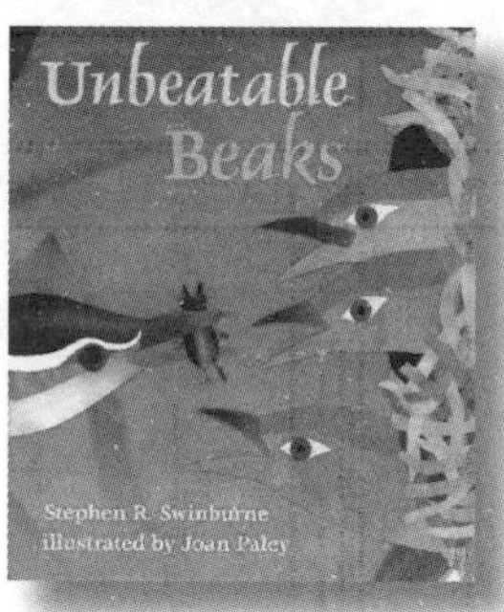

TITLE: ***Unbeatable Beaks***
AUTHOR: **Stephen R. Swinburne**
ILLUSTRATOR: **Joan Paley**
PUBLISHER: **Henry Holt**
YEAR: **1999**
GENRE: **Non-Narrative Information**
SUMMARY: *Rhyming verses describe many types of bird beaks. Includes factual information about 39 birds.*

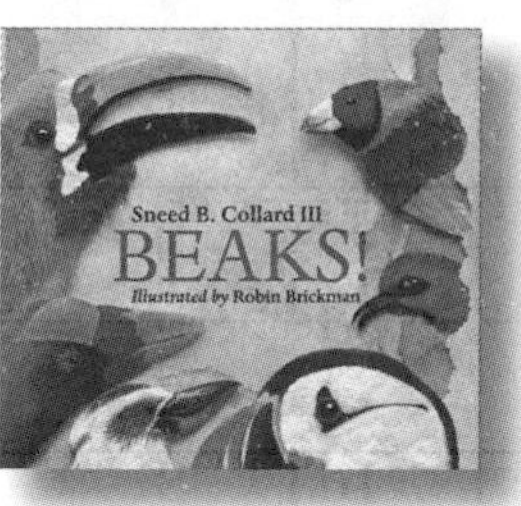

TITLE: ***Beaks!***
AUTHOR: **Sneed B. Collard III**
ILLUSTRATOR: **Robin Brickman**
PUBLISHER: **Charlesbridge**
YEAR: **2002**
GENRE: **Non-Narrative Information**
SUMMARY: *Simple text describes various bird beaks and how birds use them to eat, hunt, and gather food.*

Time Needed

This lesson will take several class periods. Suggested scheduling is as follows:

A few weeks in advance: **Engage** with Feeder Watch

Day 1: **Engage** with *Unbeatable Beaks* Read-Aloud and **Explore** with What's the Best Beak? Check-Mark Lab

Day 2: **Explain** with *Beaks!* Read-Aloud and **Elaborate** with Bird Picture Sort

Day 3: **Evaluate** with The Right Beak for the Job

Materials

For Feeder Watch (per class)

- A variety of bird feeders, including a hummingbird feeder (seven do-it-yourself bird feeder plans can be found at *http://earth911.com/news/2012/04/03/diy-recycled-bird-feeders)*
- Different types of feed (sunflower, thistle, suet, corn, fruit, nectar, mealworms, and so on)
- (Optional) Baby monitor
- (Optional) Poster of common feeder birds (available at *www.birds.cornell.edu/pfw/FreeDownloads.htm)*

NOTE: You will need an outdoor area for the bird feeders.

For What's the Best Beak? Check-Mark Lab

Per group of four students

- 1 metal pie pan
- Cup of miniature marshmallows
- ½ cup of large dry beans
- Cup of water

Per student

- Fork with the tips of the tines taped together to make a strainer
- Spring clothespin
- Eyedropper
- Small bowl
- 3 small paper cups
- Marker

SAFETY

Check with the school nurse regarding student medical issues (e.g., allergies to tree nuts or peanuts that might be present in bird feed) and how to deal with them.

For Bird Picture Sort (per pair)

- Bird Cards set

Student Pages

- What's the Best Beak?
- Bird Cards
- The Right Beak for the Job

Background

Our modern lifestyle limits children's exposure to and therefore their connection with the natural world. Observing birds at feeders is a simple way to bring nature into the classroom and help children establish a connection with their natural surroundings. This activity is also helpful in teaching about the relationship between animals' structures and their functions. *A Framework for K–12 Science Education* suggests that the relationship between structure and function in living things is a concept that students begin to understand in the early grades. In grades K–2, this involves students noticing that living things have external parts that they use in different ways to help them survive. A bird's beak is a good example of one of these external parts, or structures. The primary function of a bird's beak is to help the bird take in food. In this lesson, students will notice the various sizes and shapes of bird beaks and relate them to the type of food the birds eat. Types of beaks include those specially adapted for plucking fruit or insects from a tree, eating seeds or grains, chiseling insects out of bark, sipping nectar, tearing meat, and filtering food from water. Beaks can also help birds build nests, groom, fight, feed their young, dig, and even attract a mate.

A bird's beak grows continuously throughout the life of the bird. Although bird beaks vary widely in size, shape, and color from one species to another, they have the same basic underlying structure—an upper jaw that is fixed to the skull and a lower jaw that can move independently. The three birds featured in this lesson—hummingbird, toucan, and flamingo—have very different beaks. The hummingbird has a long, thin beak for sipping nectar. Different-shape hummingbird beaks fit different flowers. Some are curved, like the white-tipped sicklebill hummingbird, and some are straight, like the ruby-throated hummingbird. The toucan has a long, light beak that is a good tool for plucking berries and insects from tree branches. The flamingo actually eats with its beak upside down. It stands in shallow lakes and marshes and draws water through its beak by using its tongue as a pump. Once in the beak, the water is pushed through special strainers that filter out tiny plants and animals.

engage

Feeder Watch

Ahead of time, set up various types of bird feeders in a place where students can observe the birds that visit throughout the day. Show students the types of food that belong in each type of feeder (sunflower, thistle, suet, corn, fruit, nectar, mealworms, and so on). Record the names of each kind of food and the corresponding feeder. You may want to set up a baby monitor near the bird feeders and turn it on during appropriate times throughout the day so that students can hear the sounds the birds make while visiting the feeders. We suggest you hang up a poster identifying common feeder birds for your area so the students can identify some of the birds. The Cornell Lab of Ornithology offers free downloads of common feeder birds posters at *www.birds.cornell.edu/pfw/FreeDownloads.htm*.

Ask

- ? What have you noticed about the classroom bird feeders?
- ? Do you have bird feeders at home?
- ? Why do people put up bird feeders? (so they can enjoy looking at and listening to birds,

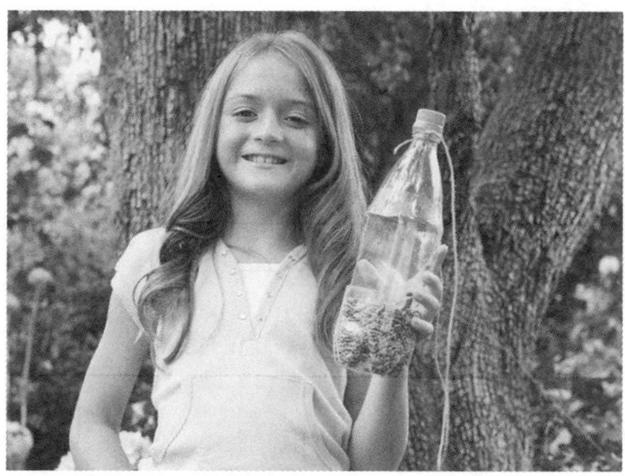

A HOMEMADE BIRD FEEDER

to help birds find food, to learn more about birds, and so on)

Model your own observations and wonderings about the birds visiting the feeders, as shown in the examples below.

Observations

- I notice that the doves usually feed on the ground instead of on the feeder.
- I have seen a lot of chickadees eating the songbird mixture.
- The goldfinches almost always go to the thistle feeder.
- There are a lot of sunflower seed shells on the ground.

Wonderings

- I wonder why certain kinds of birds go to the same feeder.
- I wonder why the sunflower seeds feeder is the first to be empty.
- I wonder why the hummingbirds are the only ones that drink from the nectar feeder.
- I wonder how the birds get the sunflower seeds out of the shells.

Unbeatable Beaks Read-Aloud

Show students the cover of the book *Unbeatable Beaks.* Tell students that the author, Stephen Swinburne, became interested in birds when he was only 9 years old. Read the second paragraph of the "Author's Note" on page 3:

> *My passion for birds started when I was nine with a pet parakeet named Magoo. I'd let Magoo perch on my finger and hold him up to my face while he pecked salt off my nose. I think it was this early face-to-face encounter with Magoo's beak that got me thinking about bird beaks.*

Making Connections: Text-to-Self

Ask students to think about each question, then share their answers with a partner:

? Have you ever had a pet bird, or known anyone who had a pet bird?

? Would you let a pet bird peck you on the nose? What would it feel like?

? What different kinds of bird beaks have you seen?

Then read the book aloud.

Rereading

> Connecting to the Common Core
> **Reading: Informational Text**
> Integration of Knowledge and Ideas: K.7, 1.7, 2.7

Tell students that you are going to reread the book for a scientific purpose. Explain that you would like them to look very carefully at the different beaks featured in the book and observe their sizes and shapes. Model this by reading the spread on pages 8 and 9 that says "Bird beaks chisel, bird beaks hook, bird beaks give birds the look" and pointing out the relationship between the text and illustrations. For example, the beak of the greater flameback illustrates the chisel-type beak and the double-crested cormorant's beak looks like a hook. Next, reread each two-page spread and invite students to make observations about the characteristics of each beak and identify which illustration goes with each description. After reading, *ask*

? Why do you think the author ends the book with the line "The world's best tool is a beak. And that's that!"? What does he mean by that? (Birds use beaks as tools for getting food.)

Tell students that they are going to get to use a variety of tools and pretend they are bird beaks to find out what's the best beak!

explore

What's the Best Beak? Check-Mark Lab

SAFETY

- Instruct students not to eat or taste any of the food items used for this part of the lesson.
- Use caution when working with pinch hazards like spring clothespins.

Pass out a pie pan containing about 1 cup of miniature marshmallows to each group of four students. Tell them that they are going to use the marshmallows to help them investigate different kinds of bird beaks. The question they are going to investigate is, "What's the best beak for picking up the marshmallows?" Give each student a fork with the tines taped together at the end, a spring clothespin, an eyedropper, a small bowl, and a copy of What's the Best Beak? student page. Tell them that after they complete each step of the lab, they should put a check mark in the box next to that step.

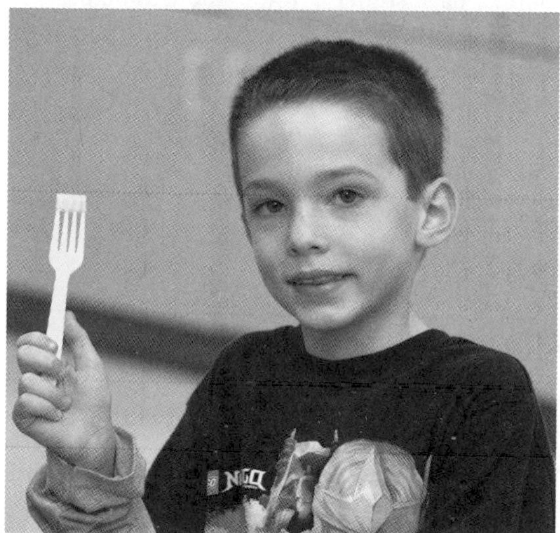

FORK STRAINER

1. Food Source: Miniature Marshmallows (berries)

Ask students to pretend that the marshmallows are berries, and that the fork, clothespin, and eyedropper are models of different kinds of bird beaks. They are not exactly like real bird beaks, but they work in similar ways. Explain that using these models will help them find out how different birds use different-shape beaks to take in food. Then have them make a prediction:

? If these tools are different kinds of bird beaks, which one do you think is best for picking up the food? Hold up the "beak" that you think would pick up the most marshmallows in 20 seconds.

? Why did you choose that beak? (Answers will vary; allow time for a few students to explain their reasoning.)

Have students circle their predictions on the What's the Best Beak? student page. Remind them that they should put a check mark in the box next to the sentence that reads, "Predict which beak will pick up the most marshmallows (circle)" after they complete that step. Then explain that they are going to count how many marshmallows they can pick up and place in their bowls with each beak. They will write down that number in the table. The rules are as follows:

1. You must keep one hand behind your back.
2. You will only have 20 seconds to get the food out of the pan and into your bowl.

Tell students that they are going to be testing each beak, beginning with the fork. Have students pick up their forks with one hand and put their other hand behind their back. When you give the signal, they should all try to pick up as many marshmallows as possible while you time them for 20 seconds. *Note:* You may want to play 20 seconds of lively music as they try to pick up the food. We like "Yakety Sax" by Boots Randolph.

After 20 seconds, have students count the number of marshmallows in their bowls and record that number in the data table. They should then return all of the marshmallows they collected to the pan for use in the next trial.

Repeat the procedure using the clothespin and then the eyedropper. *Ask*

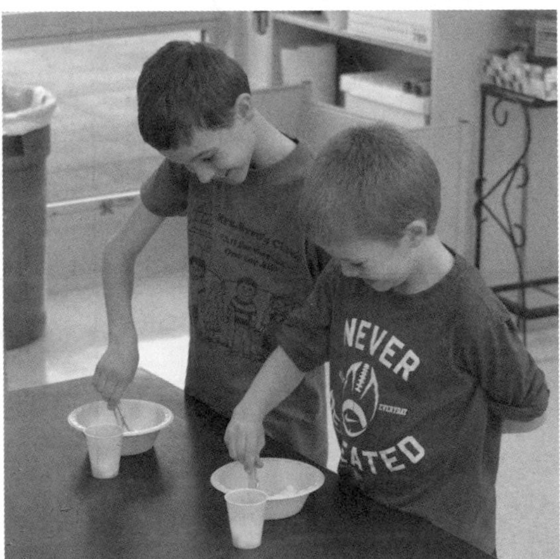

What's the Best Beak? Check-Mark Lab

? What's the best beak? Which tool worked best for picking up the marshmallows? (The clothespin should work best, although some students may find the fork worked better than the clothespin.) Have students compare their results with others.

2. Food Source: Beans in Water (tiny animals in water)

Remove the marshmallows from the pans and bowls and pour 1 cup of water and ½ cup of dried beans into each group's pie pan. Tell students that the beans represent tiny animals swimming in the water near the shoreline and, as before, the tools are models of different kinds of bird beaks. Have them predict which beak they think would pick up the most beans in water, and repeat the procedure outlined above. Then *ask*

? What's the best beak? Which tool worked best for picking up the beans in the water? (The fork "strainer" should work best, although some students may find the clothespin worked better than the fork.) Have students compare their results with others.

3. Food Source: Water (nectar)

Tell students that the water remaining in the pan represents the sweet nectar inside of flowers. Some types of birds use their beaks to get nectar to eat. This time, students will use each tool to try to get the water out of the pan and into a cup. Pass out three small paper cups to each student. Instead of counting, students will dump the water they collected with each tool into a separate cup. They should label the cups before starting: "F" for fork, "C" for clothespin, and "E" for eyedropper.

Have students predict, then test all three beaks and compare the amount of water in each labeled cup. They should circle the beak that picked up the most water on their student page. *Ask*

? What's the best beak? Which tool worked best for picking up the water? (The eyedropper should work best.) Have students compare their results with others.

4. Match each tool to a bird's beak.

After all the beaks have been tested on all the food sources, have students think about the kinds of beaks that the tools might resemble. They can draw a line from the tool to their best guess for the bird with that kind of beak. Tell them that they will find out the answers later, when you read a book called *Beaks!*

explain

Discussion

Ask

? What is the best beak: the fork, the clothespin, or the eyedropper? (It depends on which food you are trying to get.)

Explain that bird's beaks are like tools that birds use to get their food. Different kinds of birds have different kinds of beaks. *Ask*

? Have you ever seen a bird that has a beak like an eyedropper?

? Have you ever seen a bird that has a beak like a clothespin?

? Have you ever seen a bird that has a beak that can strain like a fork?

Beaks! Read-Aloud

Determining Importance

> Connecting to the Common Core
> **Reading: Informational Text**
> Integration of Knowledge and Ideas: K.7, 1.7, 2.7
> Key Ideas and Details: K.2, 1.2, 2.2

Show students the cover of *Beaks!* and tell them that book can help them learn more about how birds use their beaks. Ask students to signal when they see an illustration or hear a description of a beak that is like the fork, clothespin, or eyedropper.

Read the large print part of the book aloud. Stop at page 7 and *ask*

? Which tool from our What's the Best Beak? lab does the hummingbird beak remind you of? (eyedropper)

Read the small print on page 7 that describes how a hummingbird uses its beak to reach the nectar it needs to survive. *Ask*

? What is it about a hummingbird's beak that makes it good for getting nectar? (It is long and thin so it can fit inside a flower.)

Continue reading the large print aloud until you get to page 10. *Ask*

? Which tool from our exploration does the toucan's beak remind you of? (clothespin)

Read the small print on page 10, which describes how a toucan uses its beak to pluck berries and insects from tree branches. *Ask*

? What is it about a toucan's beak that makes it good for plucking berries and insects from trees? (It is big and strong and can snap shut to pluck food from trees.)

Continue reading the large print until you get to page 13. *Ask*

? Which tool from our exploration does the flamingo's beak remind you of? (strainer fork)

Read the small print on page 13, which describes how a flamingo uses its beak to strain tiny plants and animals out of the water. *Ask*

? What is it about a flamingo's beak that makes it good for straining tiny plants and animals out of the water? (It has a strong tongue for pumping in water and special strainers to filter out the tiny plants and animals in the water.)

Read the rest of the large print in the book aloud to students, including the beak quiz on pages 30 and 31. (Answers are on page 32.)

After reading, *ask*

Finding a Beak Like an Eyedropper

? What do you think the author of this book, Sneed Collard, is trying to tell us about bird beaks? In other words, what is the main topic of this book? (The shape of a bird's beak is related to what the bird eats.)

Making Connections

Refer back to the bird feeders students have been observing outside of the classroom. Ask questions about your specific feeders and local birds, such as

? Why do we usually only see hummingbirds at the nectar feeder? (Their beaks are good at sipping nectar.)

? Why do so many goldfinches and house finches go to the thistle feeder? (Their beaks are good at eating small seeds.)

? Why do the woodpeckers go to the suet feeders? (Their beaks are good at pecking and pulling things out.)

elaborate

Bird Picture Sort

Tell the students that they will be practicing their observation skills using bird picture cards. Pass out a set of bird picture cards to each pair of students, and ask the students to sort the bird pictures based on the type of beak each has. Then have students give each group a label to describe the shapes of the beaks, and have them hypothesize what those birds might eat and why.

After students have had the opportunity to sort and label, *ask*

? What characteristics did you use to sort the birds? (shape and/or size of beak)

? How did you decide what the birds in each group might eat? (Birds with small beaks probably eat small things like seeds, birds with sharp beaks probably eat other animals, and birds with long bills are probably used to filter small plants and animals from the water.)

? Were there birds that were hard to sort? Why?

Using a set of bird picture cards, show each of the following categories and the birds that belong in them. Allow students to re-sort their cards into these groups as you discuss each one.

Grain/Seed Eating

- Mourning dove
- House sparrow
- Chickadee
- Cardinal
- Nuthatch
- Goldfinch

Chiseling

- Woodpecker

Nectar Feeding

- Hummingbird

Meat Eating

- Hawk
- Eagle

Filter Feeding

- Mallard duck
- Canada goose

After students have sorted the birds into the correct categories, *ask*

? Why do we need different kinds of feeders with different kinds of food at our class feeders? (Different beaks are better at getting different types of food. The more variety of food we have, the more different kinds of birds will visit our feeders.)

evaluate

The Right Beak for the Job

Give each student a copy of The Right Beak for the Job student page. Have each student choose a bird from *Unbeatable Beaks* or *Beaks!* or a favorite bird to write about and draw. Have them fill in the blanks in the sentence, "A ________________ has a beak that is right for ____________ because it __________________." Then have students draw

Inquiry Place

The students can continue to observe bird feeders and brainstorm questions about birds. Examples of such questions include

? How many different bird species can we observe at our classroom feeders? Identify them!

? Which birds do we see most often? Does this change with the seasons? Tally them!

? Which birds from our area migrate during the winter? Research them!

Then have students select a question to investigate or research as a class, or have groups of students vote on the question they want to investigate or research as a team. Have students present this information on a poster. Students will then share their findings in a poster session or gallery walk. You can also report your bird sightings by joining Project FeederWatch through the Cornell Lab of Ornithology: *www.birds.cornell.edu/pfw*

a picture that illustrates the sentence. For example, they might write something like "A bald eagle has a beak that's right for tearing meat because it is sharp and curved" and draw a picture of a bald eagle tearing meat. (This is only one of several correct responses.)

Websites

Audubon Just for Kids
http://web4.audubon.org/educate/kids

Cornell Lab of Ornithology (home page)
www.birds.cornell.edu

Cornell Lab of Ornithology Project FeederWatch
www.birds.cornell.edu/pfw

Cornell Lab of Ornithology All About Birds
www.allaboutbirds.org/guide/search.aspx

More Books to Read

Collard, S. B. 2008. *Wings*. Watertown, MA: Charlesbridge.
Summary: Birds are not the only animals with wings. Bats and some insects have them, too! This book explores the diversity of these fascinating structures as well as the ways they help animals survive.

Jenkins, S., and R. Page. 2003. *What Do You Do With a Tail Like This?* Boston: Houghton Mifflin.
Summary: With Steve Jenkins's colorful collage illustrations, Page explains how different animal structures serve different functions to help the animal survive.

Pallotta, J. 1989. *The Bird Alphabet Book*. Watertown, MA: Charlesbridge.
Summary: Alphabet books are not just for learning the alphabet! Children will learn more about familiar and exotic birds through this familiar format.

What's the Best Beak?

1. **Food Source: Mini Marshmallows (berries)**

☐ Predict which beak will pick up the most marshmallows (circle):

Fork Clothespin Eyedropper

☐ Count how many marshmallows you pick up with each beak in 20 seconds, and write the number in the table below.

☐ Circle the beak that picked up the most marshmallows.

"Beak"	Number of Marshmallows
Fork	
Clothespin	
Eyedropper	

2. **Food Source: Beans in Water (tiny animals in water)**

☐ Predict which beak will pick up the most beans (circle):

Fork Clothespin Eyedropper

☐ Count how many beans you pick up with each beak in 20 seconds, and write the number in the table below.

☐ Circle the beak that picked up the most beans.

"Beak"	Number of Beans
Fork	
Clothespin	
Eyedropper	

3. **Food Source: Water (nectar)**

- ☐ Predict which beak will pick up the most water (circle):

Fork Clothespin Eyedropper

- ☐ Label three cups: "F" for Fork, "C" for Clothespin, and "E" for Eyedropper
- ☐ Put the water you pick up in 20 seconds with each beak into the right cup.
- ☐ Circle the beak that picked up the most water:

Fork Clothespin Eyedropper

4. **Match each tool to a bird's beak.**

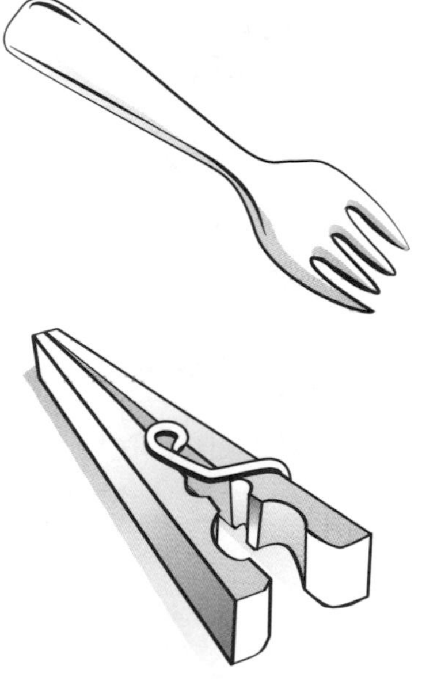

Hummingbird

Toucan

Flamingo

Bird Cards

Woodpecker

Mourning Dove

House Sparrow

Nuthatch

Chickadee

Mallard Duck

Chapter
13

Bird Cards

Hummingbird

Cardinal

Eagle

Goldfinch

Hawk

Canada
Goose

Name ______________________________

The Right Beak for the Job

A ______________________________ has a beak

that is right for ______________________________

because it ______________________________.

Chapter 14

Ducks Don't Get Wet

Description

Why does water roll off of a duck's back? How can ducks spend so much time swimming in water, even when it is cold? In this lesson, students discover the answer to these questions as they learn about the many structures that serve various functions in a duck's growth, survival, and behavior.

Suggested Grade Levels: 3–5

LESSON OBJECTIVES *Connecting to the Framework*

LIFE SCIENCES

Core Idea LS1: From Molecules to Organisms: Structures and Processes

LS1.A: Structure and Function

By the end of grade 5: Plants and animals have both internal and external structures that serve various functions in growth, survival, behavior, and reproduction.

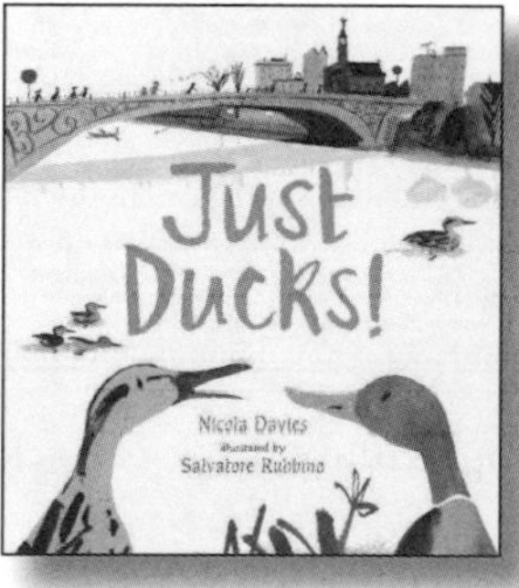

Featured Picture Books 

TITLE: ***Just Ducks!***
AUTHOR: **Nicola Davies**
ILLUSTRATOR: **Salvatore Rubbino**
PUBLISHER: **Candlewick Press**
YEAR: **2012**
GENRE: **Dual Purpose**
SUMMARY: *This dual-purpose book tells the story of a girl watching and listening to the ducks near her home. It also presents information about the structures and behaviors of ducks.*

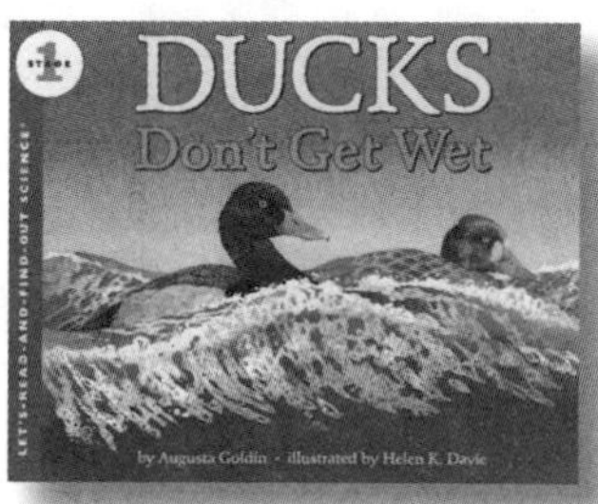

TITLE: ***Ducks Don't Get Wet***
AUTHOR: **Augusta Goldin**
ILLUSTRATOR: **Helen K. Davie**
PUBLISHER: **HarperCollins Publishers**
YEAR: **1999**
GENRE: **Narrative Information**
SUMMARY: *This* Lets-Read-And-Find-Out Science *book explores the process of preening along with other behaviors and structures ducks use to survive in their environments.*

Time Needed

This lesson will take several class periods. Suggested scheduling is as follows:

Day 1: Engage with *Just Ducks!* Read-Aloud and **Explore** with Duck Observations

Day 2: Explain with Duck Journal Discussion and **Elaborate** with Oil and Water

Day 3: Explain with *Ducks Don't Get Wet* Read-Aloud

Day 4: Elaborate with Preening and **Evaluate** with Why Don't Ducks Get Wet?

Materials

For Duck Observations (per student)

- Colored pencils

For Oil and Water (per group of four students)

- Clear plastic water bottles or plastic containers with screw-on lids, half full of water
- Small disposable cup one-third full of vegetable oil
- Food coloring

For Preening (per group of four students)

- Shallow bowl of vegetable oil
- Spray bottle of water
- Two feathers (available at craft stores)
- Paper towels

Student Pages

- My Duck Journal (Copy cover back-to-back with pages 1 and 6. Copy pp. 3 and 4 back-to-back with pp. 5 and 2.)
- Oil and Water (data sheet)

Background

A Framework for K–12 Science Education states: "A central feature of life is that organisms grow, reproduce, and die. They have characteristic structures (anatomy and morphology), functions (molecular-scale processes to organism-level physiology), and behaviors (neurobiology and, for some animal species, psychology)" (p. 143). The *Framework* recommends that children in grades 3–5 learn that plants and animals have both internal and external structures that serve various functions in growth, survival, behavior, and reproduction. Understanding structure and function can be challenging for elementary students. Many students hold the misconception that individual organisms develop structures deliberately in response to their environments. However, organisms don't develop structures because they "want to" or "try to." Biological adaptation involves naturally occurring variations in populations, not individuals. Structures do not occur overnight; rather, they take many generations of surviving in similar environments. For example, the ancestors of macaws (large parrots) may not have always had large beaks, but over long periods of time the birds with the larger beaks were able to crack open nuts better or tear apart fruits better than their smaller-beaked relatives, and thus live longer. Those birds were then able to breed more and pass their genetic traits to their offspring.

This lesson focuses on the observable structures and behaviors of ducks that help them survive in their environment, rather than how those structures and behaviors have evolved over time. Ducks have many structures and behaviors that children can readily observe. Students might notice ducks using their bills to spread oil over their feathers, coating them with a waterproof seal. This behavior is called *preening*. At the base of their tails, ducks have a structure called an *uropygial gland*, or oil gland, which produces uropygial oil; this is a waxlike substance that helps a bird waterproof its feathers. Without this protective barrier, a duck's feathers would quickly become waterlogged while swimming and the duck would become chilled. An adult duck's feathers are so waterproof that even when the duck dives underwater, its downy underlayer of feathers will stay completely dry. Ducklings, covered in down feathers, rely on their mothers for this oil until their own oil glands are fully developed. Preening behavior not only helps ducks spread oil to protect themselves from water but also helps them align and repair their feathers and remove parasites. Another important structure in ducks is the presence of webbing between the toes, which enables them to paddle when floating and propel themselves underwater when diving. So why don't ducks' feet freeze, even though they aren't covered in a layer of fat or feathers? Ducks have structures to prevent this from happening—a specialized heat exchange system between the arteries and veins in their legs and feet. This blood exchange system keeps heat loss from the duck's body to a minimum while preventing frostbite in the feet.

Although all ducks are omnivores, most ducks can be categorized into two broad groups based on their feeding behaviors: *dabbling,* or puddle, ducks and *diving* ducks. Dabbling is a behavior in which ducks nibble at the surface of the water with their beaks to get tiny bits of food. Dabbling ducks typically feed in shallower water and often *upend*, a behavior in which they push their heads under the surface of the water to feed with their tails raised in the air. Dabblers walk well and feed readily on land, and only rarely dive. Diving ducks are agile swimmers and dive well below the surface of the water in search of food. Some of the diving ducks also engage in dabbling behavior. Diving ducks do not walk as well on land as dabbling ducks because their legs tend to be located farther back on their bodies to help propel them when diving underwater. Duck bills are highly specialized structures that come in a wide variety of shapes and sizes, depending on the food sources of a particular species. Some dabbling ducks, such as the ubiquitous mallards, have a long bill with a hard nail-like structure at the tip of the bill that helps with foraging, and comb-like structures called *lamellae* on the sides of the bill that function like sieves to strain food from water. Some diving ducks, such as mergansers, have serrated toothlike lamellae for grasping fish. Both books in this lesson feature mallard ducks, which are perhaps the most familiar of all ducks. They are found all over Europe, North America, and Asia, as well as in Australia and New Zealand.

engage

Just Ducks! Read-Aloud

Connecting to the Common Core
Reading: Informational Text
CRAFT AND STRUCTURE: 3.4, 4.4, 5.4, 3.5, 4.5, 5.5

Show the students the cover of *Just Ducks!* and introduce the author and illustrator.

Making Connections

Ask

? Have you ever seen ducks swimming in a lake or pond? If so, where? Have students share their experiences with a partner.

Open the book to pages 6 and 7 where the story begins. Show students that this book has two kinds of words. The larger-print text tells the story and the smaller-print text shares facts about ducks that are not part of the story. Tell them that the first time through, you are just going to read the large words. Read *Just Ducks!* aloud to the class. After reading, *ask*

? If ducks spend so much time in the water, how do you think they keep from getting too cold or wet?

? What happens when you spend a lot of time in water? (Students may say that they get cold or their skin becomes wrinkled.)

explore

Duck Observations

Writing

Connecting to the Common Core
Writing
RESEARCH TO BUILD AND PRESENT KNOWLEDGE: 3.7, 4.7, 5.7

Tell students that the author of *Just Ducks!* Nicola Davies, loves ducks. She says, "Ducks are so lovable. As a very little girl, I was always asking to be taken to the pond in the park to go look at them. My first job was at a sanctuary for wild geese and ducks, and I spent many hours trying to draw their lovely shapes and colors."

Explain to students that *ornithologists*—people who study birds—spend a lot of time observing them. They look and listen carefully, draw pictures, record their behaviors, and write down their ideas and questions about the birds. Tell students that you are going to give them an opportunity to observe ducks just like ornithologists would. *Ask*

- What would be the best place to look for ducks? (a watery place like a lake or pond)

Although it is ideal to have students make actual observations of live ducks for this lesson, we realize that many schools do not have access to a pond or lake where ducks can be observed. Instead, students can make these observations by watching some duck video footage (see "Websites" section of this lesson). Give each student a copy of the Duck Journal and a pack of colored pencils and tell them that as they observe the ducks they should make a detailed drawing of a duck (p. 1), record the duck's behaviors (p. 2), and write down questions they have about ducks in the "Duck Wonderings" section (p. 3). It is a good idea to model some of your own questions about ducks during this time so that students hear some good examples, such as "Why do they put their heads under the water? Why do they look like they are biting their tail feathers?" Whether you are watching the video or observing ducks outdoors, it is important that during the observation time the class is quiet and still so all students can hear and see what the ducks are doing without any distractions. If you use the videos, you may need to play them more than once so students have time to observe and write.

explain

Duck Journal Discussion

After the Duck Observation activity, have students share the Duck Drawing section of their journals with a partner. *Ask*

? What structures, or parts of the duck's body, help it survive in its habitat? (Webbed feet make it easier to swim, wings make it able to fly, bright colors help the males to attract a mate, dull colors make it harder for predators to see the females, the shape of the bill helps them to find and eat certain kinds of food, and so on.)

Next ask individual students to share some of the duck behaviors they recorded. Make a list of these behaviors on the board and discuss how those behaviors might help the duck survive in its habitat.

Tell students that you are going to read the smaller-print text from the book, which might explain some of these behaviors and answer some of their wonderings. Read this text aloud, stopping periodically to discuss important vocabulary, such as *dabbling, drakes, predators, preening,* and *upending.* After reading, *ask*

? Why do you think the author included the smaller-print text sections in the book in addition to the story? (so the reader can learn important facts about ducks)

? Why do you think she chose to make this print smaller than the larger text that tells the story? (so the reader would know that the facts are not part of the story the girl is telling)

Explain to students that this type of book is called a dual-purpose book. It has two purposes: (1) telling a story and (2) giving information. The author and illustrator decided to distinguish between these two purposes by using two different sizes and styles of text.

After reading, refer back to your chart on the board and apply the correct terminology to the behaviors the students observed, such as *dabbling, upending,* and *preening. Ask*

OBSERVING OIL AND WATER

? How do you think these behaviors help the duck to survive in its habitat? (Dabbling helps them get small insects and seeds off the surface of the water, upending allows them to eat underwater plants and animals, diving helps them find food that is deep underwater, sitting on the nest keeps the eggs warm, quacking loudly calls other ducks to join them or helps alert the flock to danger, and so on.)

SAFETY

Be careful to quickly wipe up any spilled water or oil on the floor. This is a slip/fall hazard, which can result in a serious injury.

elaborate

Oil and Water

Tell students that you would like to take some time to learn more about the preening process and how it helps ducks survive. Model how to use the index in the back of *Just Ducks!* to find the word "preening" in the text. Then read the explanation about preening on page 9: "When they preen, ducks spread oil from a little spot just under their tails all over their feathers to keep them shiny and waterproof." *Ask*

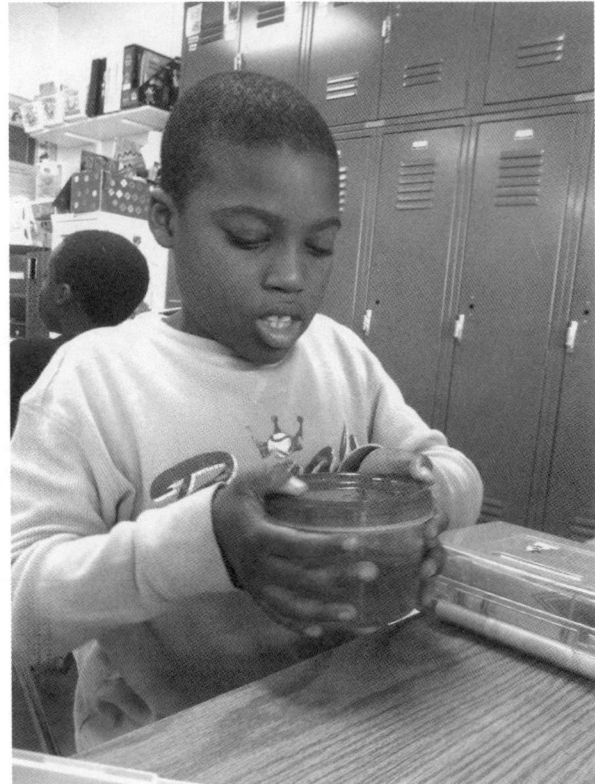

SHAKING THE OIL AND WATER

? Did anyone observe a duck preening?

Tell students that to fully understand why ducks preen, they need to learn some things about oil and water. Divide the students into groups of four and pass out the Oil and Water investigation data sheets. Give each group a plastic container half full of water and a bottle of food coloring. Instruct each group to put two drops of food coloring in their container, reseal the lid and shake it. Have the students draw and record their observations. Then give each group a small disposable cup one-third full of vegetable oil. Instruct the groups to pour the oil into their container, reseal the lid, and shake it once again. Have the students record their observations right away.

Have students place the container on the table and watch carefully to see what happens to the oil-and-water mixture. Wait approximately two minutes and then have the students record their observations. They will see the oil and water separating, with the oil on top and colored water on the bottom. *Ask*

? Do you think you can shake it hard enough that the oil and water stay mixed together?

Try it! Challenge students to shake the container hard enough and long enough so that the oil will mix with the water. Give the class two minutes to try and dissolve the oil in the water, and then ask them to record their findings. (You may want to play some fun music while they are shaking the containers.)

Ask

? What did you notice when you put food coloring in the water? (The water changed color.)

? What did you notice when you put oil in the container? (Students may have noticed that the oil floated.)

? What happened when you shook the container and waited a few minutes? (The oil and water separated.)

? Could you shake the container enough to get the oil to dissolve? (No, it is not possible to shake it hard enough to make the oil dissolve in the water.)

? So how does preening, or coating its feathers with oil, affect a duck in the water? (The water won't mix with the oil on the duck's feathers.)

explain

Ducks Don't Get Wet Read-Aloud

Determining Importance

> Connecting to the Common Core
> **Reading: Informational Text**
> KEY IDEAS AND DETAILS: 3.1, 4.1, 5.1

Introduce the book *Ducks Don't Get Wet.* Tell students that the book includes information about structures and behaviors that help ducks survive. *Ask*

? What is the difference between a structure and a behavior? (A structure is an actual body part and a behavior is something the animal does.)

Ask students to signal when they hear about a structure (part of the duck's body) or behavior (activity) that helps a duck survive in its environment. Read the book aloud, stopping to ask students if the thing they are signaling about is a structure or a behavior.

After reading, ask students to recall some of the structures (body parts) mentioned in the book that help ducks survive. Explain that the purpose of each structure and behavior is known as its function. Have students turn to page 4 of their Duck Journals and fill in the T-chart, listing structures in the left-hand column and functions (how they help the duck survive) in the right-hand column.

Possible correct responses are shown in the sample T-chart.

Structure	Function
Oil gland	Provides oil that is used for preening
Webbed feet	Help ducks swim to get food and get away from predators
Bills	Help ducks pull plants and other food from the water and helps the duck spread oil on its feathers
Feathers	Keep ducks warm and help it fly
Wings	Help ducks fly from place to place and get away from predators

Next, ask students to recall some of the behaviors from the book that help ducks survive. Have them turn to page 5 of their Duck Journals and fill in the T-chart, listing behaviors in the left-hand column and how they help the duck survive in the right-hand column. Possible correct responses are shown in the sample T-chart.

Behavior	Function
Preening	Keeps ducks dry and warm
Upending	Helps them find and eat food
Dabbling	Helps them find and eat food
Shoveling and straining	Helps them find and eat food
Diving	Helps them find food deeper in the water
Flying	Helps them get away from predators
Flying in a V	Helps them to fly farther without getting too tired
Migrating	Helps them find food in a warmer place in winter

elaborate

Preening

SAFETY

- Be careful to quickly wipe up any spilled water or oil on the floor. This is a slip/fall hazard, which can result in a serious injury.
- Have students wash their hands with soap and water upon completing the activity.

Tell the class that they are going to further explore the preening behavior to see just how it works. Give each group two feathers and place shallow bowls of vegetable oil and spray bottles filled with water in front of them. Tell students to lay one of the feathers flat on their table, spray it once or twice with water, and then observe it. *Ask*

? What did you observe? (The water soaked into the feather.)

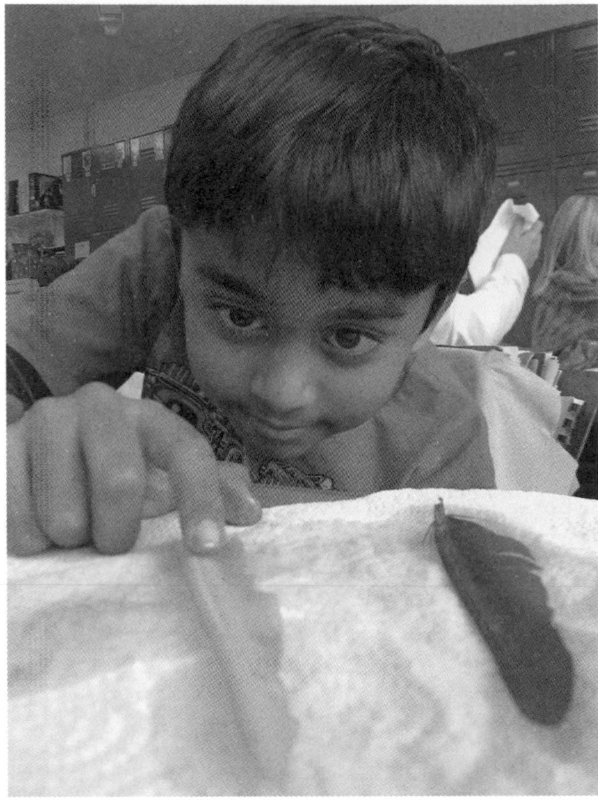

COMPARING THE FEATHERS

Tell students to preen the second feather by dipping their fingers in the oil and pulling the feather through their fingers two or three times. Once the feather is coated in oil, have students lay the feather flat on their table and spray the feather once or twice using the water bottle. *Ask*

? What did you observe? (The water did not soak into the feather. It formed little balls on the oil.)

? How does the feather that was coated with oil compare with the feather that was not coated with oil? (The water soaked into the first feather, but it did not soak into the oil-coated feather.)

Synthesizing

Ask

? How does this activity relate to something we read about in the book? (Students should realize that coating the feather with oil is similar to preening. Ducks don't get wet because they coat their feathers with oil.)

evaluate

Why Don't Ducks Get Wet?

Writing

Connecting to the Common Core
Writing
RANGE OF WRITING: 3.10, 4.10, 5.10

Have students turn to page 6 of their Duck Journals and answer the question, Why don't ducks get wet? Evaluate whether they write an accurate explanation using all of the following vocabulary: *structure, behavior, oil, water, feathers, preen, oil gland.* For example, "Ducks have a special *structure* called an *oil gland* near their tails. They *preen,* which is a *behavior* where they spread *oil* on their *feathers*. This *oil* protects the *feathers* from *water* because *oil* and *water* do not mix."

Inquiry Place

Have students brainstorm questions about ducks. Examples of such questions include

- ? How often do ducks preen themselves? Tally it!
- ? What kinds of ducks live in your area? Do they migrate? Where? Research it!
- ? How do ducks find their way when they migrate? Research it!
- ? Do all birds preen like ducks do? Do all birds have oil glands? Research it!

Then have students select a question to investigate or research as a class, or have groups of students vote on the question they want to investigate or research as a team. Have students present this information on a poster. Students will then share their findings in a poster session or gallery walk.

Websites

Cornell Lab of Ornithology
www.birds.cornell.edu

Duck Videos
www.youtube.com/watch?v=KHfE-p5jcy4
www.youtube.com/watch?v=RGDbg5QHX7Y

More Books to Read

Bancroft, H., and R. G. Van Gelder. 1996. *Animals in winter.* New York: HarperCollins.
Summary: This *Let's-Read-and-Find-Out Science* book explores the many ways animals adapt to survive winter. Hibernation, migration, and food storage are among the adaptations discussed.

Kalman, B., and Walker, K. 2000. *How do animals adapt?* New York: Crabtree.
Summary: This book explores the many ways animals adapt in order to live in their particular habitat.

Petrie, C. 2006. *Why ducks do that: 40 distinctive waterfowl behaviors explained and photographed.* Minocqua, WI: Willow Creek Press.
Summary: This reference book answers many of the common and not-so-common questions about the waterfowl species that share our world. This book would be a great resource to answer additional questions about duck behaviors and structures.

My Duck Journal
Name:
Date:

Duck Wonderings

What do you wonder about ducks? List your questions below.

Page 5

Structures and Functions

In the T-chart below, list some duck structures and their functions.

Structures	Functions

Page 2

Duck Behavior

What are the ducks doing? Describe their behaviors below.

Page 3

Behaviors and Functions

In the T-chart below, list some duck behaviors and their functions.

Behaviors	Functions

Page 4

Duck Drawing

Draw and color a duck in the box below. Show as much detail as possible and label any of the structures (body parts) you can identify.

Page 1

Why don't ducks get wet?

Explain why ducks don't get wet. Be sure to use the following words in your answer: structure, behavior, oil, water, feathers, preen, and oil gland.

Page 6

Name: ______________________

Oil and Water

1. Add 2 drops of food coloring to your container. Put the lid back on. Then shake the container.

   What do you observe?

   ____

   ____

   ____

Draw what you see

2. Add oil to your container. Put the lid back on. Then shake the container.

   What do you observe?

   ____

   ____

   ____

Draw what you see

STOP

3. Observe your container for 2 minutes. Record what happens.

____

____

____

Draw what you see

4. Do you think you can shake it hard enough that the oil and water stay mixed together? YES or NO

Shake it as hard as you can for 2 minutes. Then watch. What do you observe?

____

____

____

Draw what you see

Amazing Caterpillars

Description

In the eyes of a child, the transformation of a caterpillar into a butterfly is nothing short of magic! In this lesson students construct their own understandings about the complete butterfly life cycle by observing, exploring, and journaling about live caterpillars and butterflies before scientific explanations and vocabulary are introduced. Then they learn that butterflies aren't the only ones that go through these dramatic changes. In fact, most insects do.

Suggested Grade Levels: K-2

LESSON OBJECTIVES *Connecting to the Framework*

LIFE SCIENCES

Core Idea LS1: From Molecules to Organisms: Structures and Processes

LS1:B: Growth and Development of Organisms

By the end of grade 2: Plants and animals have predictable characteristics at different stages of development. Plants and animals grow and change. Adult plants and animals can have young. In many kinds of animals, parents and the offspring themselves engage in behaviors that help the offspring to survive.

Featured Picture Books

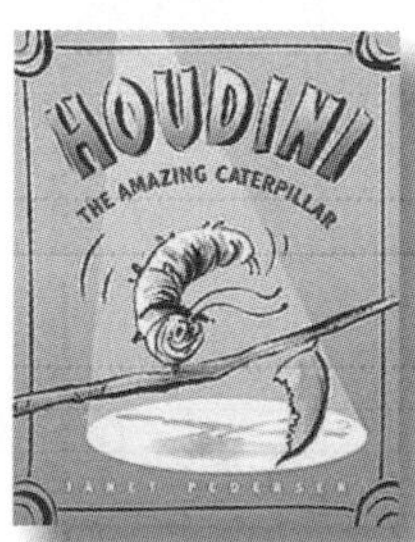

TITLE: ***Houdini The Amazing Caterpillar***
AUTHOR: **Janet Pedersen**
ILLUSTRATOR: **Janet Pedersen**
PUBLISHER: **Clarion Books**
YEAR: **2008**
GENRE: **Story**
SUMMARY: *Houdini is a classroom caterpillar with "magical" abilities. He makes a leaf vanish, crawls across a high-wire stick, and grows before your very eyes. When the audience tires of his tricks, he performs one final act to amaze everyone, including himself.*

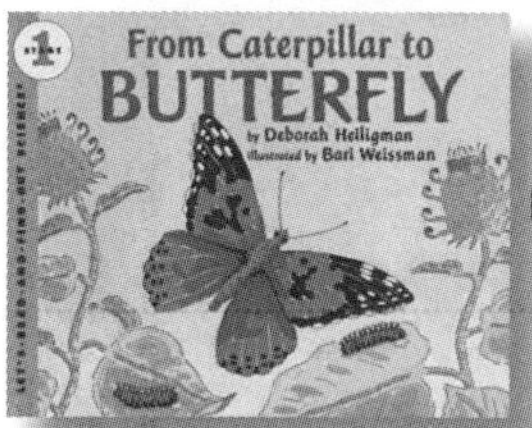

TITLE: ***From Caterpillar to Butterfly***
AUTHOR: **Deborah Heiligman**
ILLUSTRATOR: **Bari Weissman**
PUBLISHER: **HarperCollins**
YEAR: **1996**
GENRE: **Narrative Information**
SUMMARY: *Simple text and colorful illustrations describe a butterfly's life cycle within an elementary classroom.*

TITLE: ***The Very Hungry Caterpillar***
AUTHOR: **Eric Carle**
ILLUSTRATOR: **Eric Carle**
PUBLISHER: **Penguin Group**
YEAR: **1969**
GENRE: **Story**
SUMMARY: *The story follows a caterpillar as it makes some unusual food choices and grows into a beautiful butterfly.*

Time Needed

This lesson will take several class periods. Suggested scheduling is as follows:

Part 1

About Two weeks: **Engage** with *Houdini the Amazing Caterpillar* Read-Aloud and **Explore** with Our Amazing Caterpillars Journal

Part 2

Day 1: **Explain** with *Houdini the Amazing Caterpillar* and *From Caterpillar to Butterfly* Read-Aloud, card sort, and Pasta Life Cycle Diagram

Day 2: **Elaborate** with Butterflies Aren't the Only Ones

Day 3 **Evaluate** with *The Very Hungry Caterpillar* Retelling Book

NOTE: Allow about two weeks between part 1 and part 2 of this lesson. During those two weeks, students will need several opportunities to observe the caterpillars/chrysalides and record these observations in their journals.

Materials

For caterpillar observations (per class)

- Painted lady butterfly kit containing caterpillars or caterpillar coupons, food, and mesh butterfly habitat. (Butterfly kits are available at *www.insectlore.com* and *www.carolina.com*.)

For card sort (one set per pair of students)

- Butterfly Life Cycle cards (cut out in advance by teacher)

For Pasta Life Cycle Diagram (per student)

- Pasta shapes to represent the stages of the butterfly life cycle: small beadlike pasta such as pearl pasta, acini di pepe, or couscous (egg), rotini (caterpillar), shell (chrysalis), and bowtie (butterfly)
- Green, brown, and black markers or colored pencils
- 9" or 10" paper plate
- White glue

Student Pages

- Our Amazing Caterpillars Journal (Copy the cover on a separate sheet, then copy pp. 1 and 8 back-to-back with 2 and 7; copy pp. 6 and 3 back-to-back with 4 and 5. Fold and staple along the spine.)
- Butterflies Aren't the Only Ones
- *The Very Hungry Caterpillar* Retelling Book

Background

A Framework for K–12 Science Education recommends that by the end of grade 2 students recognize that plants and animals have predictable characteristics at different stages of development. With young children, it is best to develop understanding of biological concepts such as this through direct experience with living things and their life cycles. Butterflies are ideal to use in this type of exploration. They are familiar, interesting, and often found in the local ecosystem.

Butterflies experience *metamorphosis* as they progress through their life cycles. Almost all insects go through metamorphosis, with the exception of a few such as silverfish and springtails (which hatch from the egg as a small replica of the adult and mature by growing bigger). Metamorphosis is marked by distinct stages, often resulting in an abrupt change of physical appearance. The butterfly begins its life as an egg and hatches into a *larva*, known as a caterpillar. In this *larval stage*, the caterpillar spends most of its time eating plant material. As it grows, the exoskeleton becomes tight, so it molts, or sheds, the old one. This process is repeated several times as the caterpillar grows. Finally, the caterpillar makes a special silk "button" which it uses to hang upside down from a twig or other surface. The soft body that the caterpillar reveals when it molts for the final time is called a *chrysalis* (plural: *chrysalides*), which hardens to form a protective shell. Many children and adults confuse this structure with a cocoon; however, a cocoon is a silk covering that most moths spin to protect their chrysalis. (Some species of moth caterpillars also disguise their cocoons with leaves or other debris.) Inside the chrysalis, a remarkable transformation takes place. This is known as the *pupal stage* of the butterfly life cycle, during which the butterfly is immobile. In the pupal stage, parts of the caterpillar's body break down into undifferentiated cells, which then form new structures. The amount of time required for this transformation varies depending on the species. After the caterpillar completely transforms, the chrysalis cracks open and the *adult* butterfly emerges. At first, its wings are damp and crumpled. As it pumps blood into them, the wings straighten out and dry. Soon the adult butterfly is ready to take its first flight.

Many animals (such as fish, birds, reptiles, and mammals) have fairly simple life cycles in which they are born or hatched, then grow into adults. Amphibians have a bit more complex life cycle where they hatch from an egg, live underwater breathing through gills, grow into adults, move to land and breathe air through their skin and lungs. Most insects, however, have a dramatic life cycle in which they go through the following stages: egg, larva, pupa, and adult. A few insects, like dragonflies, grasshoppers, and cockroaches, go through an incomplete metamorphosis, which means they go from egg to larva to adult, without a pupal stage. In the elaborate phase of this lesson, students apply what they learn about the stages of insect metamorphosis to other insects, so they will understand that butterflies aren't the only ones that go through this dramatic change.

About the Painted Lady Butterfly

This lesson uses painted lady butterflies to demonstrate the stages of a metamorphic life cycle. The painted lady butterfly (*Vanessa cardui*) is found in most of the United States as well as Canada. Several companies have marketed ready-made kits that include everything you will need to help your butterflies develop. Most contain coupons to send away for the live caterpillars. You are unlikely to see the egg stage of the life cycle; therefore, you should be prepared to describe this to your students. Female painted lady butterflies lay their eggs on plants such as thistle, hollyhock, and mallow. When the caterpillars hatch, they eat the leaves of the plants. The eggs are light blue-green in color and the size of a pinhead. The caterpillars emerge within three to five days.

Excluding the egg stage, the life cycle should take approximately three weeks. Your caterpillars will likely ship in a ventilated plastic cup containing food (referred to as "nutrient" by some vendors). The caterpillars will spend most of their time eating and may spin a small amount of silk. You will notice small balls in the container; the balls are caterpillar waste, or *frass*.

When your caterpillars are ready for the pupal stage, they will crawl to the lid of the container and hang upside down in a hook shape. If they are not disturbed, the caterpillars will harden into chrysalides in one to two days. Once all of your caterpillars have entered the pupal stage, you should remove the lid and hang it inside of your butterfly home. You may notice the chrysalides shaking. This is natural and helps to prevent predator attacks in the wild.

The adult butterflies will emerge in 7 to 10 days. You may notice that the chrysalides become transparent just before this occurs. The new butterflies' wings will be wrinkled and soft. They will be ready to fly within a few hours. Do not be alarmed if you notice your butterflies excreting a red liquid. This is meconium, a waste product from the pupal stage. You may want to line the bottom of your butterfly home with paper towels or a paper plate to catch the meconium.

To feed your butterflies, dissolve 3 teaspoons of sugar into 1 cup of water. Sprinkle this mixture onto fresh flowers or saturate a crumpled paper towel. Place them in your butterflies' home. You will see your butterflies uncurl their proboscises and sip the sugar water.

When you are ready to release your butterflies, make sure the temperature is above 55°F. Painted lady butterflies are one of the most widely distributed butterfly species in the world, but check your state's wildlife release rules prior to setting them free in your area. If you are unable to release them, your butterflies may mate and lay eggs. When the larvae hatch, add thistle, hollyhock, or mallow leaves to your butterflies' home. The adult butterflies have a life span of only two weeks. Reassure your students that this, too, is a part of the life cycle, and the young will continue the circle of life.

engage

Houdini the Amazing Caterpillar Read-Aloud

Making Connections: Text-to-World

Ask

- ? Have you ever seen a magic show?
- ? What kinds of tricks do magicians do? (make something disappear, pull a rabbit out of a hat, guess a card, escape from an enclosed space, and so on)
- ? Have you ever heard of a magician named Houdini?

Tell the students that Harry Houdini was one of the most well-known magicians of all time. Houdini became famous for his escape acts. He could free himself from prisons, handcuffs, ropes, chains, and water-filled containers. One of Houdini's most famous illusions involved making a full-grown elephant and its trainer disappear from the stage! Then tell the students that you have a book about a very unusual magician, also named Houdini, to share with them. Show the students the cover of *Houdini the Amazing Caterpillar,* and introduce the author and illustrator.

Inferring

Connecting to the Common Core
Reading: Literature
Key Ideas and Details: K.1, 1.2, 1.3

Read the book aloud, stopping after page 19. ("Suddenly, he knew just what his next act would be. It would be his most daring act ever.") *Ask*

? How are Houdini's actions like magic? (Houdini makes leaves vanish by eating them; he grows before their very eyes and breaks free from his skin; he holds a pose for almost two weeks without food, water or taking a break.)

Model questioning by saying, "I wonder what Houdini's next act will be?" Give each student a sticky note, and ask them to write or draw what they think Houdini's next, most daring act will be. Collect the sticky notes and use them to assess student preconceptions. Tell students that they are going to see some "magic" in their own classroom before you read the rest of the book to them.

explore

Our Amazing Caterpillars Journal

Connecting to the Common Core
Writing
Research to Build and Present Knowledge: K.7, 1.7, 2.7

Hold up the caterpillar container, and tell students that you have some amazing little "Houdinis" for them to observe.

Pass out the Our Amazing Caterpillars Journals. Tell students they are going to be observing these caterpillars for several weeks, just as the children in *Houdini the Amazing Caterpillar* did. For observation 1, first have students record the date. Throughout the day, allow small groups of students to visit the caterpillars to record their observations. Continue to have the students make observations every few days, or as needed when changes occur. At this point, it is not necessary to reveal the scientific terminology for each stage, although some students may already be familiar with some of these words. You will continue with the rest of this lesson when all of the caterpillars have become adult butterflies.

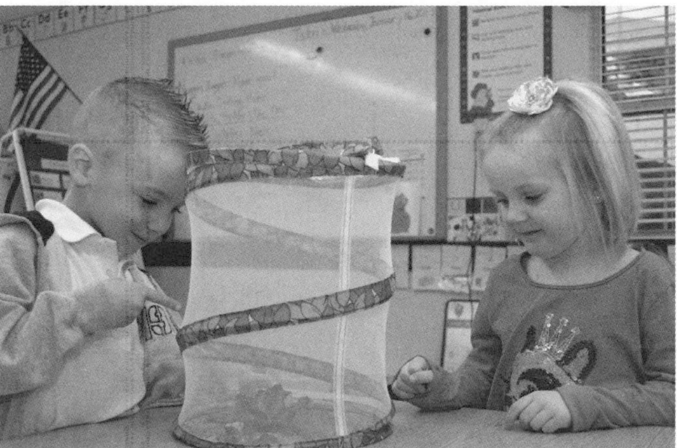

Observing the Butterfly Life Cycle

Use the following guidelines to help students work through the journal:

Observation 2: Complete one to two days following the initial observation.

Observation 3: Complete two to three days following the second observation.

Observation 4: Complete when at least one caterpillar is hanging upside down, preparing for the pupal stage.

Observation 5: Complete once all chrysalides have formed, and you have removed them from the caterpillar container. Allow students to make their observations before

Painted Lady Chrysalis

placing the chrysalides in the butterfly home. The drawing of the circle is the disk upon which the chrysalides are attached.

Observation 6: Complete when one or more butterflies have emerged. This can also be used if you are lucky enough to observe a butterfly as it is emerging.

Observation 7: Complete once all of the butterflies have emerged.

Wonderings: Students can record any questions they have about the caterpillars and butterflies on this page.

explain

Houdini the Amazing Caterpillar and *From Caterpillar to Butterfly* Read-Aloud

Making Connections: Text-to-World

Tell students that now that they have witnessed the "magic" of their own caterpillars, you will read the entire story of *Houdini the Amazing Caterpillar* to them. Tell them that as you read, you would like them to signal when they see Houdini doing something that they saw their caterpillars doing. Students should signal that both Houdini and their classroom caterpillars ate, shed their skin, grew, hung upside down for a long time, and emerged as butterflies. After reading, *ask*

? Is this book fiction or nonfiction? How do you know? (fiction, because the caterpillar smiles and talks)

Tell students that you have a nonfiction book that can help them learn more about the life cycle of a butterfly, including the names of all the stages that they observed. Introduce the author and illustrator of *From Caterpillar to Butterfly.* As you show the cover, *ask*

? What do you think this book has in common with Houdini the Amazing Caterpillar? (Answers may include that they are both about butterflies, there are caterpillars in both books, and the butterflies in both books are orange.)

Open Sort

Before reading the nonfiction book, give each pair of students a set of precut Butterfly Life Cycle cards with the names and pictures of the four stages of butterfly metamorphosis: egg, caterpillar (larva), chrysalis (pupa), butterfly. Before reading, ask students to put the cards in the order in which they think they go.

Determining Importance

> Connecting to the Common Core
> **Reading: Informational Text**
> Key Ideas and Details: K.3, 1.3, 2.3

Tell students that as you read the book aloud, you would like them to listen for the names of the stages on their cards. Read the large-print text of the book aloud. As you read about each stage, have students move their cards in the correct order. Be sure to read the small-print text on page 11 ("The caterpillar is also called the larva.") and page 19 ("The chrysalis is also called the pupa.").

Closed Sort

After reading, make sure students have put the cards showing the stages of the butterfly life cycle in the correct order: egg, caterpillar (larva), chrysalis (pupa), butterfly. *Ask*

? Where do butterfly eggs come from? (adult butterflies)

Have students move the cards to create a circle that represents the continuation of the egg-caterpillar-chrysalis-butterfly cycle.

Ask if students have ever heard the term *cocoon.* Explain that butterflies do not make cocoons, but moths do. Cocoons are spun out of silk during a moth's pupal stage. Some moths also camouflage their cocoons by covering them with bits of leaves or other plant material. They go through their metamorphosis inside the safety of their cocoon. *Ask*

? "Where do butterflies go through their pupal stage?" (inside a chrysalis)

Pasta Life Cycle Diagram

Pass out one paper plate and one piece of each of the following pastas to each student: beadlike pasta, rotini, shell, and bowtie. Students will also need green, brown, and black markers or colored pencils. Guide students through the following steps to make a butterfly life cycle diagram on a paper plate:

> **SAFETY**
>
> Tell students not to eat or taste any food (e.g., pasta) or beverage that has been made or used in the lab or classroom activity unless instructed to do so by the teacher.

1. *Ask*

   ? How does a caterpillar begin its life? (as an egg) Show the illustration of butterfly eggs on page 9 of *From Caterpillar to Butterfly*. Tell students to hold up the pasta that looks most like a butterfly egg (beadlike pasta). Tell them to place the beadlike pasta at the 12:00 position on their plates. Then have them sketch the outline of a leaf with a green marker or colored pencil around the "egg." Tell them that they will glue the pasta onto the "leaf" later. Have students write "egg" below the beadlike pasta.

2. *Ask*

   ? What hatches from a butterfly egg? (caterpillar) Show the illustration of a caterpillar on page 15 of the book. Tell students to hold up the pasta that looks most like a caterpillar (rotini). Tell them to place the rotini at the 3:00 position on their plates. Remind them that they will glue the pasta onto the plate later. Have students write "caterpillar" below the rotini.

3. *Ask*

   ? What is another name for a caterpillar? (larva). Have students write "larva" in parentheses next to the word "caterpillar."

4. *Ask*

   ? What happens after the caterpillar molts for the last time? (It makes a chrysalis.) Show the illustration of the chrysalis on page 20 of the book. Tell students to hold up the pasta that looks most like a chrysalis (shell pasta). Tell them to place the shell pasta at the 6:00 position on their plates. Then have them use a brown marker or colored pencil to draw a twig just above the shell pasta (so the "chrysalis" has a place to attach). Remind them that they will glue the pasta onto the paper plate later. Have students write "chrysalis" below the shell pasta.

5. *Ask*

   ? What is another name for the chrysalis? (pupa) Have students write "pupa" in parentheses next to the word "chrysalis."

6. *Ask*

   ? What comes out of the chrysalis? (butterfly) Show the illustration of the adult butterfly on page 29 of the book. Tell students to hold up the pasta that looks most like a butterfly

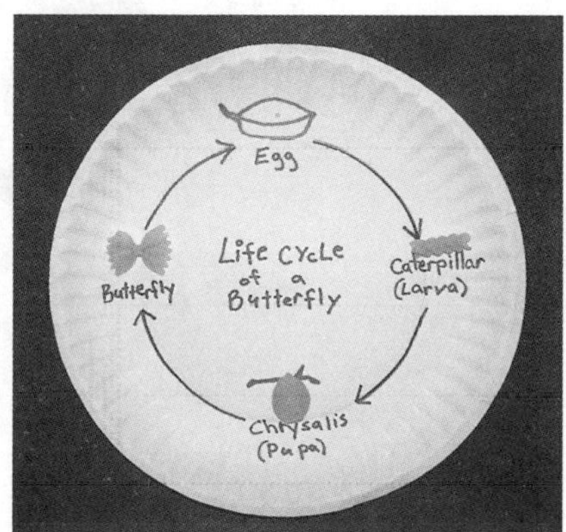

PASTA LIFE CYCLE DIAGRAM

(bowtie pasta). Tell them to place a piece of bowtie pasta at the 9:00 position on their plates. Remind them that they will glue the pasta onto the plate later. Have students write "butterfly" below the bowtie pasta.

7. Have students draw arched arrows from the egg to the caterpillar to the butterfly to show that these stages occur in a set order. *Ask*

   ? Should we draw an arrow from the butterfly back to the egg?

   Reread pages 28 and 29 of *From Caterpillar to Butterfly*. Introduce the term *life cycle* by telling students that every living thing is born, grows and changes, reproduces by laying eggs or having babies, and dies. The babies, or offspring, grow up to repeat these stages. *Ask*

   ? What is the word for the changes that a butterfly goes through in its life cycle? (metamorphosis, p. 6).

   Tell students that the word *metamorphosis* means "really big change." Explain that all living things go through a life cycle, but only some of them, like the butterfly, go through a metamorphosis. Then have students draw an arrow from the butterfly to the egg to demonstrate that the life cycle will continue.

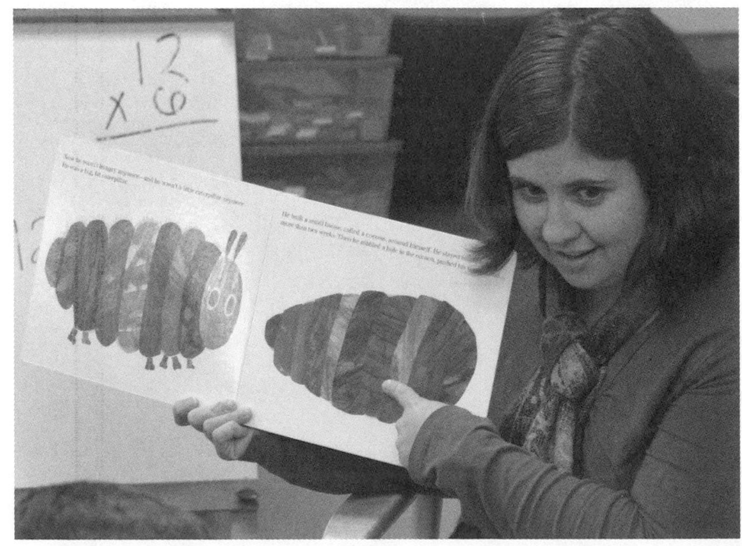

POINTING OUT SCIENTIFIC INACCURACIES

8. Have students write "Life Cycle of a Butterfly" in the center of their plates with a black marker.
9. Have students use white glue to affix the pasta shapes to their appropriate places on the paper plates.

elaborate

Butterflies Aren't the Only Ones

Explain that many animals, such as fish, birds, reptiles, and mammals, have fairly simple life cycles in which they are born or hatched, then grow into adults. Amphibians, such as frogs, have a bit more complex life cycle—they are born, live underwater breathing through gills, grow into adults, then move to the land and breathe air. Most insects, however, have a dramatic life cycle where they go through the following stages: egg, larva, pupa, and adult. Write the names of these stages on the board. *Ask*

? What is the larval stage of the butterfly called? (caterpillar)

? What is the pupal stage of the butterfly called? (chrysalis)

Explain that a few insects, like dragonflies, grasshoppers, and cockroaches, go through an incomplete metamorphosis, which means they go from egg to larva to adult, without a pupal stage. However, most insects, like butterflies, go through a metamorphosis, or really big change.

Pass out the Butterflies Aren't the Only Ones student page to each student or pair of students. Explain that they will apply the concepts and vocabulary they have learned about the stages of the butterfly life cycle to another kind of insect. Assign each student or pair of students an insect that goes through a metamorphosis and have them research the life stages of their insect on the internet or in nonfiction books. Examples of such insects include ant, bee, beetle, firefly, housefly, ladybug, mosquito, moth, praying mantis, and wasp. Have students use the student page to draw the different stages of that insect's life cycle, then have them share what they learned about their insect with the class.

evaluate

The Very Hungry Caterpillar Retelling Book

Introduce the author and illustrator of *The Very Hungry Caterpillar.* Read the book aloud to students. After reading, *ask*

? Is this book fiction or nonfiction? (fiction) How do you know? (The caterpillar eats food that caterpillars don't really eat, and so on)

Rereading and Determining Importance

Connecting to the Common Core
Reading: Literature
Craft and Structure: K.5, 1.5

Tell students that you are going to read the book again, but this time you would like to point out everything in the book that is not factual based on the things they have observed and read about caterpillars and butterflies. If they see a picture or hear something in the book they think is incorrect, they should raise their hands and explain their reasoning. Responses might include

- Caterpillars would not eat the foods in the book like ice cream, salami, cake, and so on.
- Butterfly caterpillars make a chrysalis, not a cocoon.
- Butterflies must let their wings dry before flying.
- Butterflies push out of the chrysalis. They don't nibble.

Many children have asked Eric Carle why the butterfly in *The Very Hungry Caterpillar* comes from a cocoon and not a chrysalis—so many, in fact, that it appears in the "Frequently Asked Questions" page on his website. This is the answer he gives:

> *Here's the scientific explanation: In most cases a butterfly does come from a chrysalis, but not all. There's a rare genus called* Parnassian, *that pupates in a cocoon. These butterflies live in the Pacific Northwest, in Siberia, and as far away as North Korea and the northern islands of Japan.*
>
> *And here's my unscientific explanation: My caterpillar is very unusual. As you know caterpillars don't eat lollipops and ice cream, so you won't find my caterpillar in any field guides. But also, when I was a small boy, my father would say, "Eric, come out of your cocoon." He meant I should open up and be receptive to the world around me. For me, it would not sound right to say, "Come out of your chrysalis." And so poetry won over science!*
>
> *Eric Carle,*
> www.eric-carle.com/q-cocoon.html

Tell students that Eric Carle did not write *The Very Hungry Caterpillar* as a science book, so it's okay if it is not scientifically accurate, or true. Anything can happen in fiction; a caterpillar can even eat ice cream and salami! But tell students they are going to get a chance to create a version of the story so that it *is* scientifically accurate. Pass out *The Very Hungry Caterpillar* Retelling Book. As you read it aloud, have the students illustrate the story and label the correct stages of the butterfly's life cycle. When students are finished, they should cut out each page separately and staple the pages together in order.

Inquiry Place

Have students brainstorm questions about animal life cycles. Examples of such questions include

- ? How does the moth life cycle compare with the butterfly life cycle? Observe it or research it! (*Note:* Life cycle kits for saturniid and hornworm moths are available from *www.carolina.com*.)
- ? How does the mealworm life cycle compare with the butterfly life cycle? Observe it or research it! (*Note:* Mealworms are available at pet stores and bait shops.)
- ? What kinds of cocoons do different moths make? Research it!
- ? Which insects go through an incomplete metamorphosis? Research it!
- ? What are the stages of an amphibian's life cycle? Research it!

Then have students select a question to investigate or research as a class, or have groups of students vote on the question they want to investigate or research as a team. Students can present their findings at a poster session or gallery walk.

Websites

The Children's Butterfly Site
www.kidsbutterfly.org

National Geographic video "Butterfly: A Life"
Watch the transformation of a monarch butterfly from egg to larva to pupa to adult in two minutes.
http://video.nationalgeographic.com/video/national-geographic-channel/all-videos/av-8520-8756/ngc-butterfly-a-life

More Books to Read

Allen, J. 2003. *Are you a butterfly?* New York: Kingfisher.
Summary: Simple text and soft illustrations help the readers imagine themselves as butterflies experiencing metamorphosis.

Bunting, E. 1999. *Butterfly house.* New York: Scholastic Press.
Summary: A young girl saves a caterpillar from a hungry jay and creates a home for it. She raises the caterpillar until it becomes a painted lady butterfly. The girl releases the butterfly, but as she grows older, painted ladies continue to flock to her garden.

Kalman, B. 2006. *The life cycle of a butterfly.* New York: Crabtree.
Summary: This scientific text clearly explains the stages of a butterfly's life cycle. Attractive photographs and clear illustrations enhance the fact-filled book.

Pashley, H., and L. Adams. 2010. *Look, ask and learn about butterflies.* New York: Little Science Books.
Summary: This encourages the reader to observe and ask questions about butterflies. Stunning, up-close photos show various butterflies at different stages in their life cycles.

Ryder, J. 1996. *Where butterflies grow.* New York: Puffin.
Summary: Strong imagery helps the readers connect to the story and experience life as a growing black swallowtail butterfly. The author includes suggestions for creating your own butterfly garden.

Slade, S. 2008. *From caterpillar to butterfly: Following the life cycle.* Minneapolis, MN: Picture Window Books.
Summary: Bright illustrations and informative text describe the life cycle of the monarch butterfly.

Our Amazing Caterpillars Journal

Scientist: ____________________

Date: From __________ to __________

Wonderings

What are you wondering about caterpillars?

Page 8

Observation #1

Date: ____________

Look closely at one of our caterpillars.
Draw and color it below.

Page 1

Observation #2

Date: ____________

What are the caterpillars doing? Record the number of caterpillars doing each activity below.

crawling ____

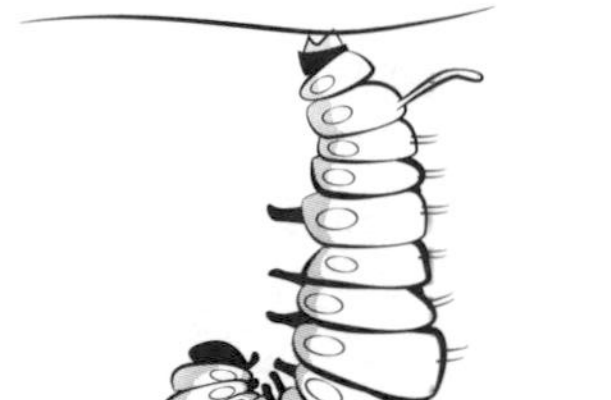

hanging ____

resting ____

eating ____

Page 2

Observation #7

Date: ____________

Draw and color what you see in the container today.

Page 7

Observation #4

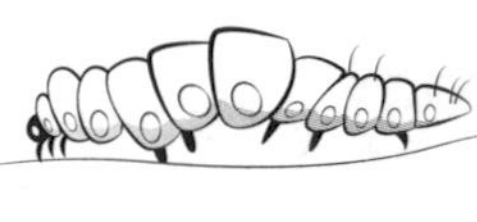

Date: ____________

Draw and color what you see in the container today.

Page 4

Observation #5

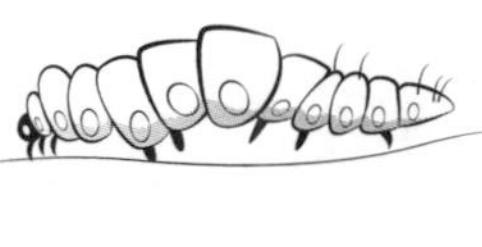

Date: ____________

Where did our caterpillars go? Draw and color what you see.

Page 5

Observation #6

Date: ____________________

What do you observe now? Draw and color what you see inside the container.

Page 6

Observation #3 

Date: ____________________

Draw and color what you see in the container today.

Page 3

Butterfly Life Cycle Cards

 Egg	 Caterpillar (Larva)	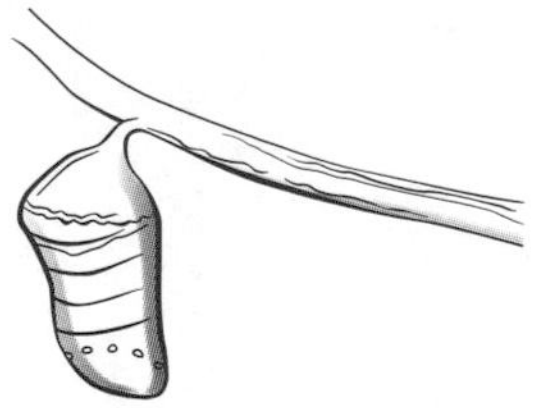 Chrysalis (Pupa)	 Butterfly
 Egg	 Caterpillar (Larva)	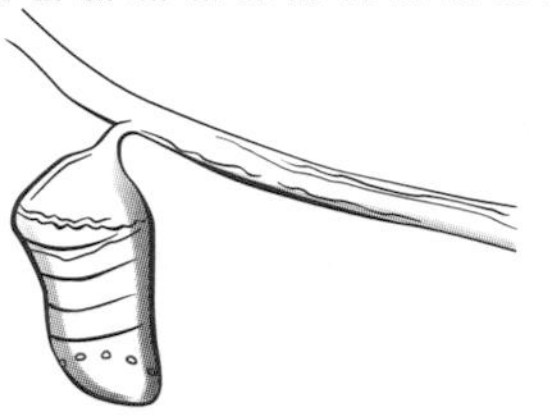 Chrysalis (Pupa)	 Butterfly
 Egg	 Caterpillar (Larva)	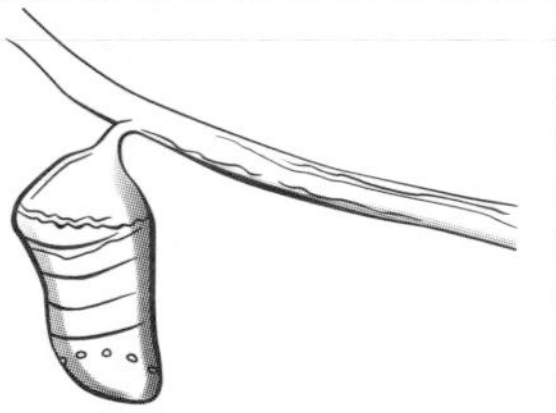 Chrysalis (Pupa)	 Butterfly
 Egg	 Caterpillar (Larva)	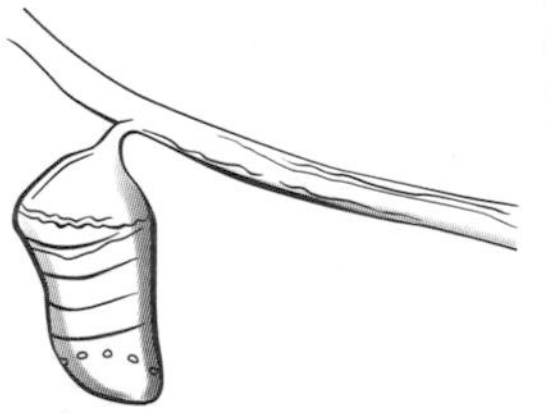 Chrysalis (Pupa)	 Butterfly
 Egg	 Caterpillar (Larva)	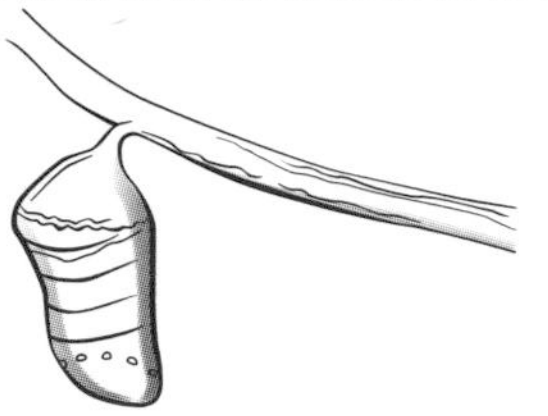 Chrysalis (Pupa)	 Butterfly
 Egg	 Caterpillar (Larva)	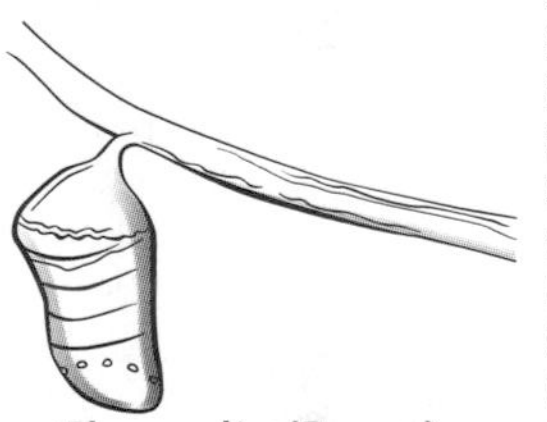 Chrysalis (Pupa)	 Butterfly

Name: ______________________________

Butterflies Aren't the Only Ones

Directions:
Draw the stages of the life cycle of an insect other than a butterfly. Be sure to draw arrows from one stage to the next.

Egg

Life Cycle of a ____________

Adult

Larva

Pupa

Name: ______________________________

The Very Hungry Caterpillar

Retelling Book

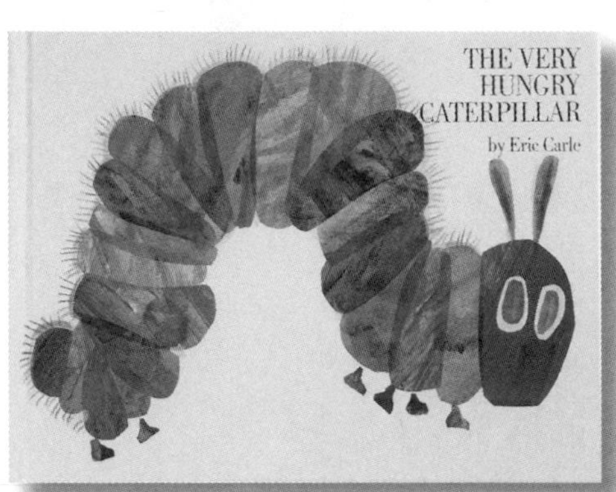

By Eric Carle

*Retold by*______________________

In the light of the moon a little egg lay on a leaf.

One warm day—pop!—a tiny and very hungry caterpillar ate its way out of the egg.

He started to look for some food.

Name: ______________________________

The Very Hungry Caterpillar

Retelling Book

For many days, the caterpillar ate through nice green leaves, and after that he felt much better.

Now he wasn't hungry any more—and he wasn't a little caterpillar any more. He was a big fat caterpillar.

He made a covering, called a chrysalis, around himself. He stayed inside for several days. Then he pushed his way out and…

he was a beautiful butterfly!

Fossils Tell of Long Ago

Description

Many young children are fascinated by dinosaurs and other prehistoric life. In this lesson, students learn that fossils are the key to learning about these once-living organisms, and that fossils not only tell us about prehistoric life but also give us clues about the Earth's environment back then. Students also discover that only a small percentage of the remains of living things actually become fossilized.

Suggested Grade Levels: 3–5

LESSON OBJECTIVES *Connecting to the Framework*

LIFE SCIENCES

CORE IDEA LS4: BIOLOGICAL EVOLUTION: UNITY AND DIVERSITY

LS4.A: EVIDENCE OF COMMON ANCESTRY AND DIVERSITY

By the end of grade 5: Fossils provide evidence about the types of organisms (both visible and microscopic) that lived long ago and also about the nature of their environments. Fossils can be compared with one another and to living organisms according to their similarities and differences.

Featured Picture Books

TITLE: ***Fossil***
AUTHOR: **Claire Ewart**
ILLUSTRATOR: **Claire Ewart**
PUBLISHER: **Walker Books for Young Readers**
YEAR: **2004**
GENRE: **Narrative Information**
SUMMARY: *A girl finds a fossil buried in the sand, which takes her imagination back in time to the life of a prehistoric pterosaur,* Ornithocheirus. *Watercolor illustrations and poetic text give readers an idea of how this ancient creature may have lived, died, and became fossilized over time.*

TITLE: ***Fossils Tell of Long Ago***
AUTHOR: **Aliki**
ILLUSTRATOR: **Aliki**
PUBLISHER: **HarperCollins**
YEAR: **1990**
GENRE: **Narrative Information**
SUMMARY: *Beginning with a description of how a prehistoric fish becomes fossilized, this book explains the process of fossilization and how scientists today learn about past life based on the fossils they find buried on Earth today.*

Time Needed

This lesson will take several class periods. Suggested scheduling is as follows:

Day 1: **Engage** with Draw a Dinosaur and *Fossil* Read-Aloud

Day 2: **Explore** with Observing Real Fossils and card sequencing and **Explain** with *Fossils Tell of Long Ago* Read-Aloud

Day 3: **Elaborate** with The Fossil Game

Day 4: **Evaluate** with *Fossil* Rereading and I Found a Fossil Writing Activity

Materials

For Draw a Dinosaur (per student)

- Blank paper or scrap paper

For Observing Real Fossils (per class)

- Animal fossils or reproductions of animal fossils

For Card Sequencing (per group of two to three students)

- Precut fossil formation cards in plastic bags

For the Fossil Game

Per class

- Die
- The Fossil Game Data Table (for teacher use, to be projected in the classroom)
- The Fossil Game Board (for teacher use)

Per student

- Precut Fossil Fortune Teller

For I Found a Fossil (per student)

- Fossil
- Colored pencils

FOR OBSERVING REAL FOSSILS

Fossil kits are available from science supply companies such as

Carolina
(*www.carolina.com*)

Nasco
(*www.enasco.com*)

Ward's Natural Science
(*www.wardsci.com*)

Student Pages

- How to Make a Fossil Fortune Teller
- Fossil Fortune Teller Templates 1, 2, and 3 (divided equally among students)
- I Found a Fossil Journal (Make book by copying cover back-to-back with pp. 1 and 6, and copying pp. 3 and 4 back-to-back with pp. 5 and 2.)
- I Found a Fossil Scoring Rubric

Background

According to *A Framework for K–12 Science Education,* children in grades 3–5 should learn that fossils are used by people today to learn about both the types of living things that lived long ago and the nature of their environments. *Paleontology* refers to the study of fossils and what fossils can tell us about the history of Earth. The American Museum of Natural History website identifies four big ideas of paleontology on the "PaleontOlogy: The Big Dig" section for kids (*www.amnh.org/explore/ology/paleontology*): (1) fossils tell stories about Earth's history, (2) fossils can't tell us everything, (3) fossils are really rare, and (4) the fossil record is like a big jigsaw puzzle, with most of the pieces missing. It is not important that students memorize the names of and specific details about different dinosaurs or prehistoric animals; rather, the focus should be on these "big ideas."

The term *fossil* refers to physical evidence of former life from prehistoric time. This prehistoric evidence includes fossilized remains of living organisms, impressions and molds of their physical form, and marks or traces created in sediment by their activities. Fossils can be divided into two broad categories: fossilized body parts (bones, teeth, skin, and so on) and fossilized traces (footprints, nests, dung, and so on). There are a variety of ways that living things can become fossilized. This lesson focuses on the process of *permineralization,* in which minerals replace the actual organic remains of the organism.

One objective for this lesson is for students to understand that it is very uncommon for living things to become fossilized. When most organisms die, they decay without a trace after natural processes recycle their soft tissues and even their hard parts such as bone and shell. For a plant or animal to become fossilized, the conditions at the time of death must be just right. If a plant or animal is not buried soon after death, fossilization often becomes impossible because of scavengers, algae, bacteria, and weather conditions such as rain, wind, water erosion, and sun exposure. This means that very few plants and animals actually become fossilized. Some scientists estimate that fewer than 2% of plant or animal species that lived on Earth have ever become fossilized.

Fossils teach us not only about the plants and animals of the past but also about the Earth's topography and climate change. Paleontologists may find tropical plant fossils in modern-day deserts or fossils of sea creatures in a modern-day farmland. These discoveries give us clues about how those places have changed over time in climate and topography. Fossils help us piece together the long history of Earth and its inhabitants.

engage

Draw a Dinosaur

Pass out blank sheets of paper or scrap paper to students. Tell the students that you want them to use a pencil to draw a picture of a dinosaur. Ask students to put as much detail in their pictures as they can. Give the students several minutes to complete this activity. Refrain from answering any questions students may ask about dinosaurs during this time, such as "How big is a dinosaur?" You want the students' thoughts to be their own impression of what a dinosaur is and looks like.

Once the students have had time to complete their drawings, have the students sit in a circle holding their drawings out for everyone to see. As the students look at each other's drawings, ask them to raise their hands if they have ever seen the dinosaur they've drawn in real life. If students do raise their hands, ask them to clarify when they saw the dinosaur. Many students may say at the museum or on TV, but point out that they have never seen a dinosaur alive with their own eyes. *Ask*

? If you've never seen the dinosaur you drew, how did you know what it looked like? (Students may say that they saw it at the museum, in a book, on the internet or on TV.)

? How do you think people who make museum displays or write books about dinosaurs know what they looked like? (They use fossils to study what the animals long ago looked like.)

Observing Fossils

Fossil Read-Aloud

Show the students the cover of the book *Fossil* and ask what they see on the cover. Explain that the animal on the cover was *Ornithocheirus* (Or-NITH-oh-KAI-rus), which was a reptile very similar to modern seafaring birds like pelicans or seagulls. Ask the students what they see drawn below the water. This is a picture of the remains of the animal. This is its fossil.

Making Connections: Text-to-Self

Tell students that as a child Claire Ewart, the author of this book, lived near a lake where she often found bits of fossils. You may want to read the jacket flap inside the back cover of the book with more information about the author. *Ask*

? Have you ever found a fossil?

Questioning

Connecting to the Common Core
Reading: Informational Text
Key Ideas and Details: 3.1, 4.1, 5.1

As you read the book aloud, including the end matter titled "Fossil Evidence," stop and *ask*

? Where did this animal live? (on an island)

? What kinds of food did this animal eat? (fish)

? What happened to *Ornithocheirus* when it died? (It sank into the silt at the bottom of the sea.)

? Why didn't its body rot (decompose)? (Lack of oxygen at the bottom of the sea protected it from completely decomposing.)

? How did the animal become fossilized? (Minerals seeped in to replace the bones.)

? Where was the animal's fossil found? (on land) How did the Earth change after the animal died? (The sea is now land.)

? What caused the fossil to be exposed? (Heat, cold, wind, sun, ice, and rain wore away the ground.)

explore

Observing Real Fossils

Pass around some real fossils for students to explore. While students are examining the fossils, *ask* guiding questions such as

? What type of animal or plant do you believe your fossil might have been? What evidence makes you think that?

? Do you think this animal or plant lived in the ocean or on land? Why do you think so?

? How do you think this fossil was formed?

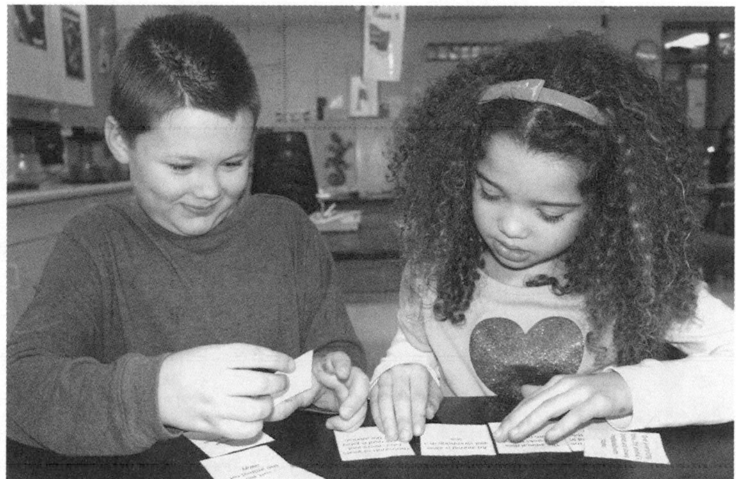

SEQUENCING THE FOSSIL CARDS

Card Sequencing (Before Reading)

Tell the students that you have a set of cards to give them. Each card has a step that a prehistoric animal may have gone through to become fossilized. Explain that they should try to put the steps in the order that they would most likely happen. Give each group a set of fossil formation cards and allow them time to put the steps in order, and then bring the class back together. Discuss how they sequenced the cards and why.

explain

Fossils Tell of Long Ago Read-Aloud

> Connecting to the Common Core
> **Reading: Informational Text**
> INTEGRATION OF KNOWLEDGE AND IDEAS: 3.9, 4.9, 5.9
> KEY IDEAS AND DETAILS: 3.3, 4.3, 5.3

Show students the cover of *Fossils Tell of Long Ago* and introduce the author and illustrator. Explain that this is a nonfiction book that will tell them more about how fossils are formed and the things that we can learn by studying fossils. As they are listening, they will get clues as to how their fossil cards should be ordered.

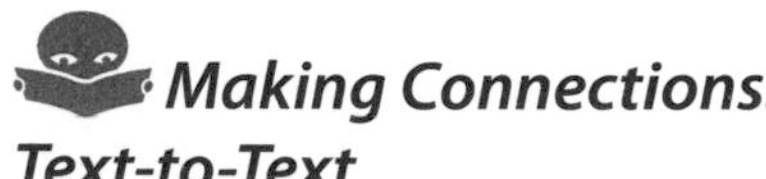

Making Connections: Text-to-Text

Read the book aloud to the class, stopping at page 12 to *ask*

? How does this book's information compare with what we read in our last book, *Fossil*? (In both stories the animal died underwater, was buried in the sand, and was found millions of years later.)

Questioning

Ask

? Do you think all fossils were buried under water? Where else might they have been buried?

Continue reading the story and stop after reading page 21 to clarify that fossils can be formed in mud, ice, stone, and amber.

Card Sequencing (After Reading)

Once you have finished reading the book aloud, have the students go back to their places with their groups and use what they have learned to reorder their fossil formation cards. Then discuss the correct order and have them rearrange their cards if necessary. The correct order is as follows:

1. The animal is alive and swimming in the sea.
2. The animal dies and sinks to the bottom of the sea.
3. The soft parts of the animal rot away.
4. The animal's bones are left on the seafloor.

Playing the Fossil Game

5. The skeleton of the animal is buried in the mud on the seafloor.
6. Over a very long time, more and more mud is piled over the animal.
7. Over a very long time, the animal's bones are slowly replaced with stone.
8. The animal becomes a fossil.
9. The fossil is discovered.

Ask

? What things can we learn from fossils? (sizes and shapes of once-living plants and animals)

? What things can't we learn from fossils? (colors or patterns of once-living plants and animals)

Explain to students that, in order to make their best guesses about what colors and patterns prehistoric animals and plants had, scientists compare them with similar animals and plants that are alive today. But no one knows for sure exactly what they looked like.

elaborate

The Fossil Game

Ask

? How common do you think it is for plants and animals to become fossilized?

Explain to the students they are going to play a game to further investigate the probability or likelihood of a plant or animal becoming fossilized. Distribute templates 1, 2, and 3 of the Fossil Fortune Teller evenly among students. Using the How to Make a Fossil Fortune Teller student page, have the students cut and fold the Fossil Fortune Tellers. You may want to read the steps together and model them for the class or make them ahead of time.

Tell the students they are going to pretend that they are a prehistoric organism that has died and that they may undergo a number of different fates: some will become fossils, while others will be eaten, washed away, or decayed over time. Say, "We will begin our game with everyone standing. If you become a fossil, you will remain standing, but if you do not become fossilized you will sit down." Next, project an image of The Fossil Game Data Table. Tell students that throughout the game you will keep track of the results using this data table.

Directions for The Fossil Game (see also The Fossil Game Board):

1. Have the students stand up and spread out in the classroom holding their assembled fortune tellers.
2. Record the number of students in the class in the "Number of Organisms" column.
3. Hold the die about an inch above the star on the game board and drop it.
4. Call out the number that lands face up, and instruct the students to open and close their fortune tellers that many times.
5. Call out the letter that the die landed on. Have the students open that panel under the corresponding letter to discover their fate. If they became a fossil, they should remain standing, otherwise they should sit down.
6. Count the number of students who are standing and record that number in the "Number of Fossils Formed" column of the data table.
7. Have everyone stand for the next round. The game ends after five rounds.

> NOTE: The purpose of this game is not to come up with a definitive fraction of organisms that become fossils; rather, it is to demonstrate the idea that most living things never become fossils because conditions aren't just right. As students read the statements on the fortune tellers they learn the reasons why.

Once the game has ended, add up the number of students who were fossilized in each round and then add up the number of students who played the game. For example, if 25 students played the game and in each round one student became fossilized, the final results would show that out of 125 organisms only 5 fossils formed.

Discuss the results of the game. Explain to the class that it is not very likely for a fossil to be formed because the conditions have to be just right. In fact, most ancient living things never became fossils. The organism had to have been preserved in ice, rock, amber, or mud quickly after its death so that it would not be eaten, washed away, or decayed.

Fossil Rereading and I Found a Fossil Writing Activity

Writing

> Connecting to the Common Core
> **Writing**
> TEXT TYPES AND PURPOSES: 3.3, 4.3, 5.3

Rereading

Reread the book *Fossil* by Claire Ewart. Have students pay special attention to the details and scientific information about *Ornithocheirus* that the author reveals through the story. Explain that this type of book, known as a narrative informational book, communicates a sequence of factual events over time within the context of a story. The author begins the story with the girl finding a fossil and then describes the life and death of the animal. Finally, she describes how it became a fossil and was discovered by the little girl. Point out that this story follows a circular pattern, beginning and ending with the girl finding the fossil. Tell students that they will be writing their own narrative information about one of the fossils they observed earlier in this lesson and they will use this circular format for their story.

Give each student a copy of the I Found a Fossil journal, an animal fossil, colored pencils, and the scoring rubric. Tell students the name of their fossils if they are not labeled. Provide resources for your students to research information about their animals such as when the animal lived, what its habitat was like at the time, and what and how it ate.

Have students share their stories with a partner or the class. Use the scoring rubric to evaluate their books.

Websites

American Museum of Natural History "Fossil Halls"
http://www.amnh.org/exhibitions/permanent-exhibitions/fossil-halls

"PaleontOlogy: The Big Dig"
www.amnh.org/explore/ology/paleontology

BBC "Science and Nature: Prehistoric Life"
www.bbc.co.uk/sn/prehistoric_life/games

More Books to Read

Aliki. 1988. *Digging up dinosaurs*. New York: HarperCollins.
Summary: This *Let's-Read-and-Find-Out Science* book gives young children a look at how dinosaurs lived millions of years ago and how scientists learn about them from the fossils they left behind.

Jenkins, S. 2005. *Prehistoric actual size*. Boston: Houghton Mifflin.
Summary: The author's trademark paper collage illustrations depict a variety of prehistoric animals

Inquiry Place

Have students brainstorm questions about fossils. Examples of such questions include

? What types of fossils, if any, can be found in the area where you live? Research it!

? Are there certain places on Earth that have more fossils than others? Why? Research it!

? What type of rock are fossils found in? Research it!

Have students select a question to investigate or research as a class, or have groups of students vote on the question they want to investigate or research as a team. You may want to direct students to the "PaleontOlogy: The Big Dig" section of the American Museum of Natural History website to do their research and come up with more questions. Students can present their findings at a poster session or gallery walk.

at their actual size. For each featured animal, the book includes some characteristics, how long ago it lived, and its length.

Kudlinski, K. 2005. *Boy, were we wrong about dinosaurs!* New York: Dutton Children's Books.
Summary: Examines what is known about dinosaur bones, behavior, and other characteristics, and explains through specific examples that our ideas about dinosaurs have changed as more evidence has been discovered.

Squire, A. 2012. *Fossils.* Chicago: Children's Press.
Summary: This book is a part of the *True Books* series that focuses on fossil formation and the information paleontologists get from the study of fossils.

Zoehfeld, K. W. 2008. *Finding the first T. rex.* New York: Random House Books for Young Readers
Summary: This book explores the work of a famous paleontologist, Barnum Brown, as he uncovered the first discovered *Tyrannosaurus rex* remains.

Fossil Formation Cards

Over a very long time, more and more mud is piled over the animal.

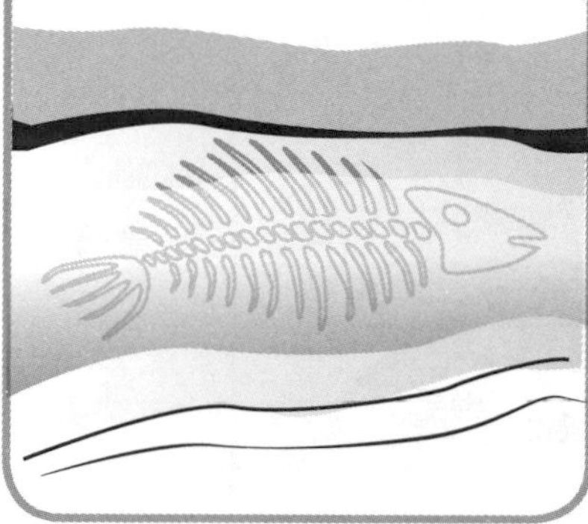

The animal dies and sinks to the bottom of the sea.

The animal becomes a fossil.

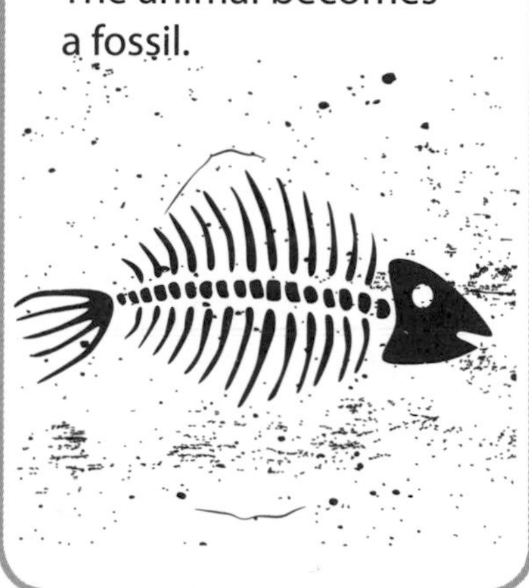

The animal's bones are left on the seafloor.

The fossil is discovered.

An animal is alive and swimming in a sea.

Over a very long time, the animal's bones are slowly replaced with stone.

The soft parts of the animal rot away.

The skeleton of the animal is buried in the mud on the seafloor.

How to Make a Fossil Fortune Teller

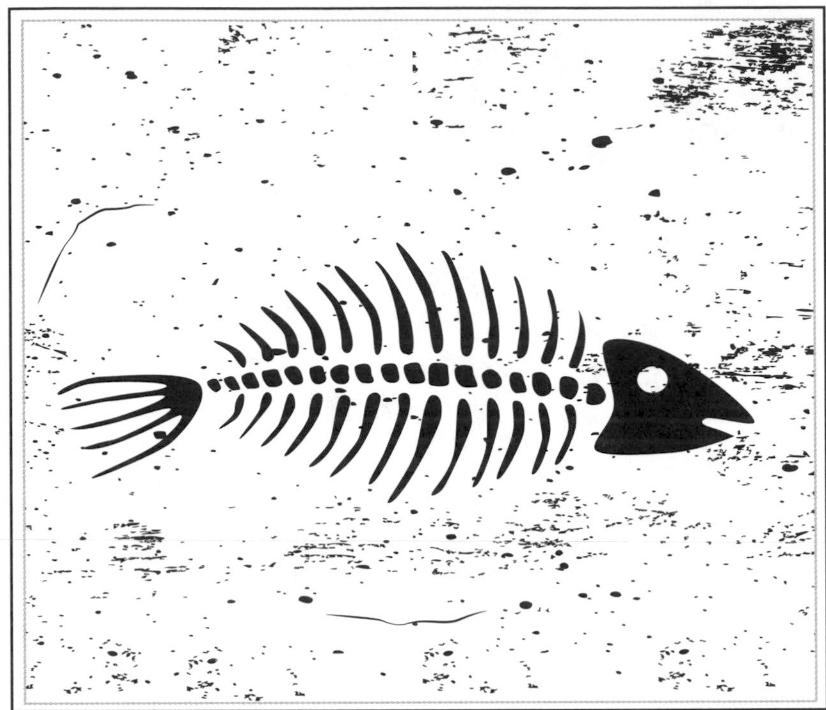

1. Cut the fortune teller template to make a square.
2. Lay the paper square with the fossil side down. Fold the square in half to make a triangle, crease it, and then open it back up.
3. Lay the paper square fossil side down again. Fold the other corners into a triangle and crease again. Unfold so that the fortune teller is a square again (fossil side down).
4. Fold each corner point into the center of the creases.
5. Flip it over (fossil side down). Fold all four corner points into the center once again.
6. Fold the square in half to make a rectangle and crease it.
7. Open it back up to the square. Fold the other way to make a rectangle and crease it.
8. Stick your thumbs and two forefingers into each of the four bone flap pockets. Fingers should press center creases so that all four flaps meet at a point in the center.

Fossil Fortune Teller

Template 1

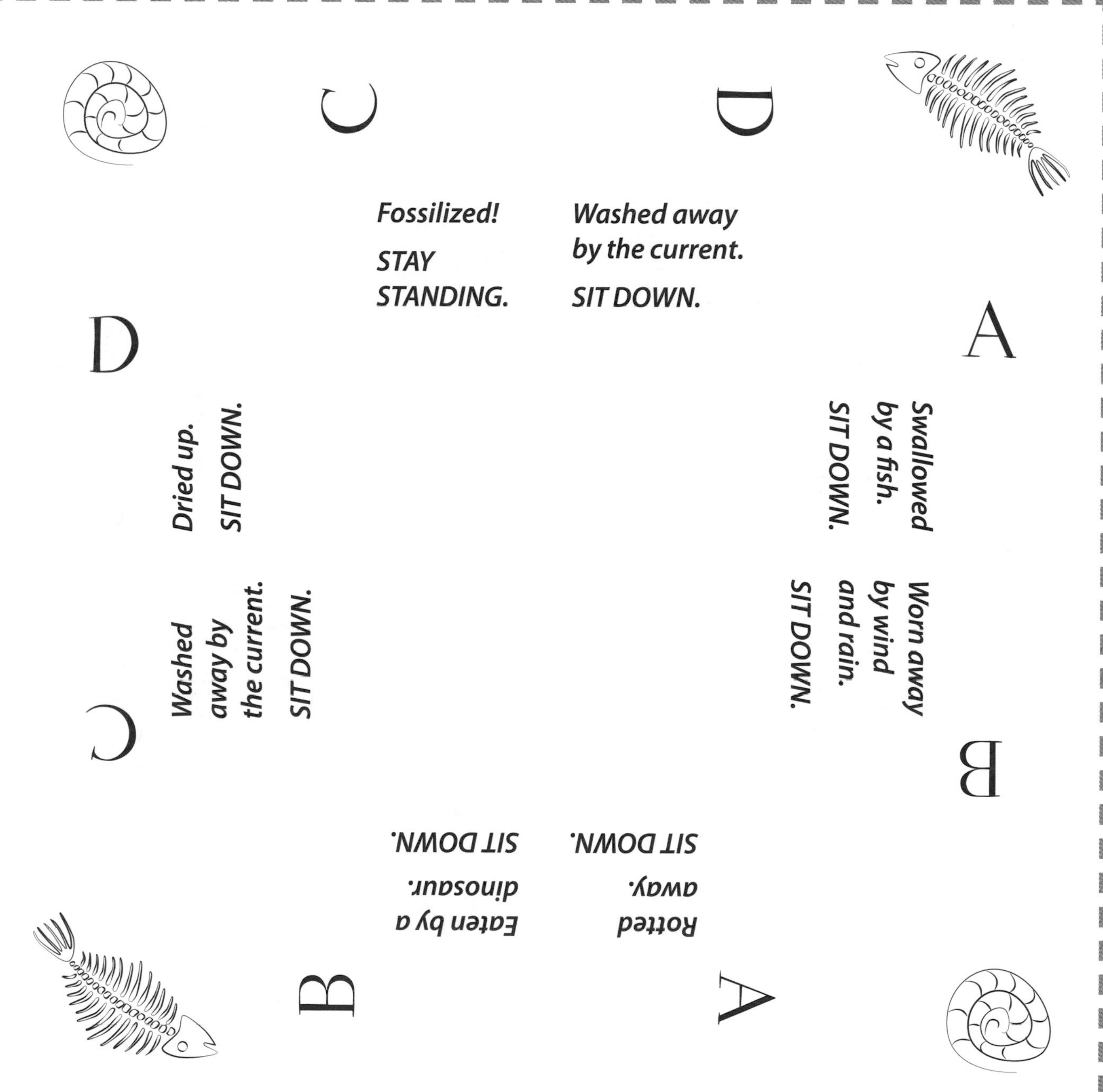

Cut off at the dotted line to make a square.

Fossil Fortune Teller

Template 2

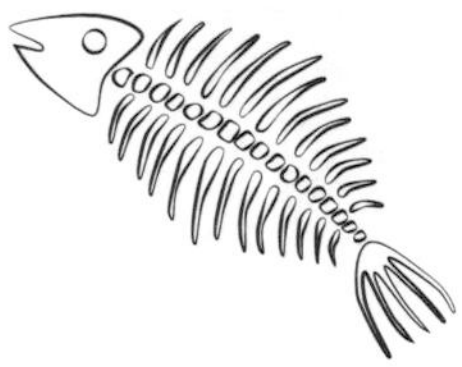

C

D

Swallowed by a fish.

SIT DOWN.

Rotted away.

SIT DOWN.

A

Rotted away.

SIT DOWN.

Eaten by an animal.

SIT DOWN.

B

Washed away by the current.

SIT DOWN.

Burned up in a lava flow.

SIT DOWN.

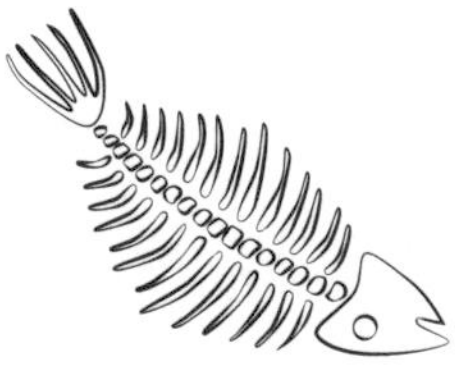

A

B

D

Fossilized!

STAY STANDING.

Dried up.

SIT DOWN.

C

Cut off at the dotted line to make a square.

Fossil Fortune Teller

Template 3

C

D

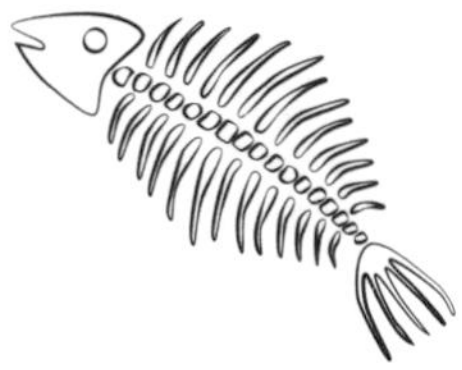

Dried up.

SIT DOWN.

Rotted away.

SIT DOWN.

D

A

Swallowed by a fish.

SIT DOWN.

Fossilized!

STAY STANDING.

Dried up.

SIT DOWN.

Eaten by a dinosaur.

SIT DOWN.

C

B

Burned up in a lava flow.

SIT DOWN.

Washed away by the current.

SIT DOWN.

B

A

Cut off at the dotted line to make a square.

The Fossil Game Board

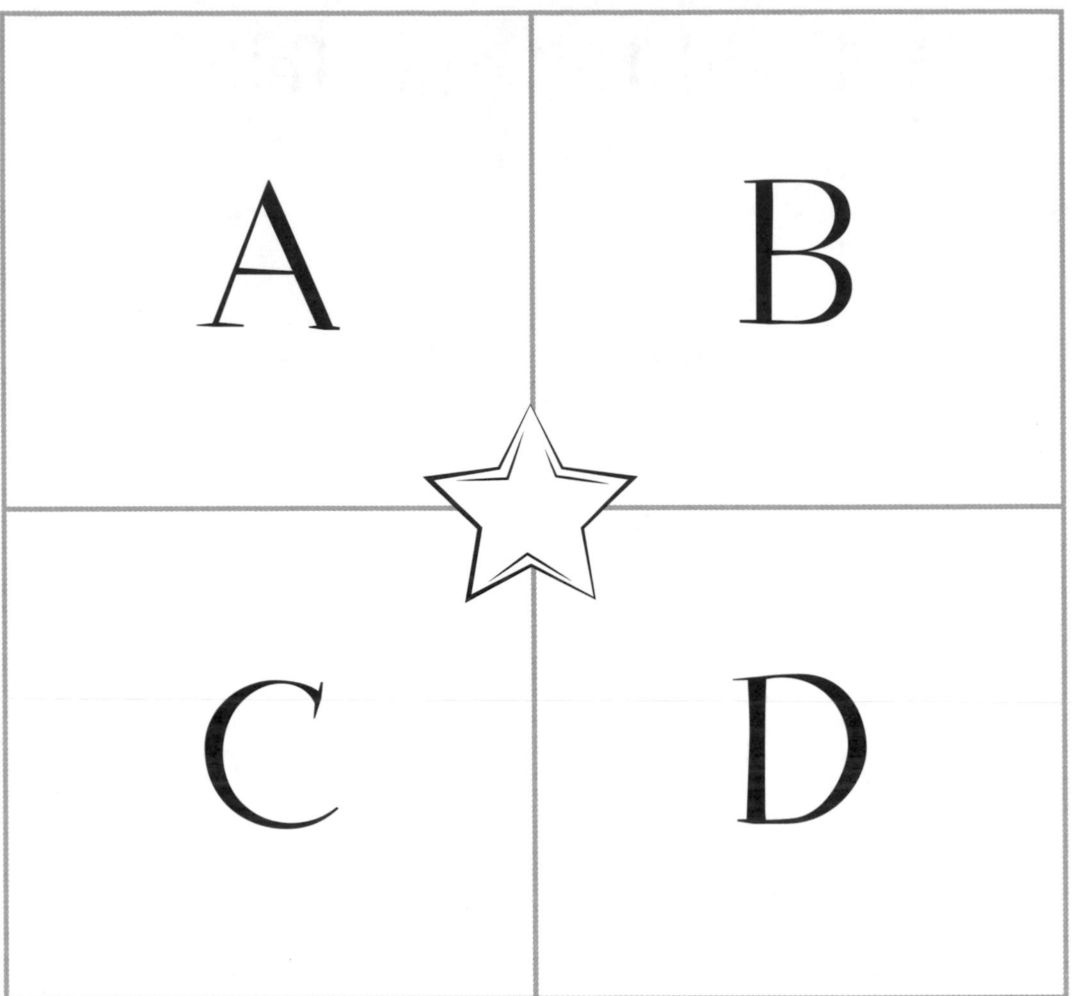

How to Play "The Fossil Game"

1. Have the students stand up and spread out in the classroom holding their assembled fortune tellers.
2. Record the number of students in the class in the "Number of Organisms" column.
3. Hold the die about an inch above the star on the game board and drop it.
4. Call out the number that lands face up, and instruct the students to open and close their fortune tellers that many times.
5. Call out the letter that the die landed on. Have the students open that panel under the corresponding letter to discover their fate. If they became a fossil they should remain standing, otherwise they should sit down.
6. Count the number of students who are standing and record that number in the "Number of Fossils Formed" column of the data table.
7. Have everyone stand for the next round. The game ends after five rounds.

The Fossil Game Data Table

Round	Number of Organisms	Number of Fossils Formed
1		
2		
3		
4		
5		
	Total Number of Organisms:	Total Number of Fossils Formed:

I found a stone that once was

____

Page 1

Page 6

I Found a Fossil

By ____________________

Page 5

Page 2

Page 4

Page 3

Name: ______________________________

I Found a Fossil

Scoring Rubric

Write and illustrate a story about finding a fossil.
Include the criteria listed below in your story.

4-Excellent **3-Above Average** **2-Average** **1-Below Average**

Score	Criteria
____ 4 ____ 3 ____ 2 ____ 1	Cover: Your name and a detailed drawing of the fossil
____ 4 ____ 3 ____ 2 ____ 1	Page 1: The name of the fossil in the sentence ("I found a stone that once was __________.") and an illustration of you discovering the fossil
____ 4 ____ 3 ____ 2 ____ 1	Page 2: A description and illustration of what the animal looked like when it was alive
____ 4 ____ 3 ____ 2 ____ 1	Pages 3 & 4: Descriptions and illustrations of the animal surviving in its habitat (What did its environment look like at the time? How did it move, get food, escape from predators? and so on)
____ 4 ____ 3 ____ 2 ____ 1	Page 5: A description and illustration of how the animal became fossilized
____ 4 ____ 3 ____ 2 ____ 1	Page 6: A description and illustration of how the fossil was uncovered and discovered by you

____ **Total Points/24**

Reduce, Reuse, Recycle!

Description

In today's world of shrinking resources and booming populations, managing solid waste is more important than ever. This lesson teaches students how we can decrease the amount of trash we send to landfills or incinerators by reducing, reusing, and recycling.

Suggested Grade Levels: K–2

LESSON OBJECTIVES *Connecting to the Framework*

EARTH AND SPACE SCIENCES

CORE IDEA ESS3: EARTH AND HUMAN ACTIVITY
ESS3.C: HUMAN IMPACTS ON EARTH SYSTEMS

By the end of grade 2: Things that people do to live comfortably can affect the world around them. But they can make choices that reduce their impacts on the land, water, air, and other living things—for example, by reducing trash through reuse and recycling.

Featured Picture Books 

TITLE: ***The Three R's: Reuse, Reduce, Recycle***
AUTHOR: **Nuria Roca**
ILLUSTRATOR: **Rosa M. Curto**
PUBLISHER: **Barron's Educational Series**
YEAR: **2007**
GENRE: **Narrative Information**
SUMMARY: *Gives examples of how the three Rs—reduce, reuse, and recycle—are used in one boy's community.*

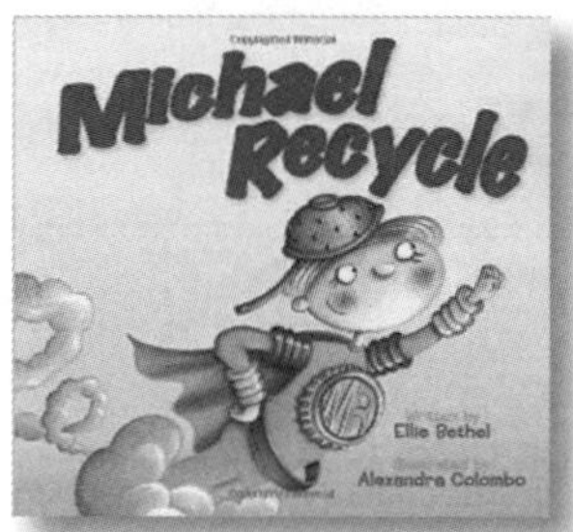

TITLE: ***Michael Recycle***
AUTHOR: **Ellie Bethel**
ILLUSTRATOR: **Alexandra Colombo**
PUBLISHER: **Worthwhile Books**
YEAR: **2008**
GENRE: **Story**
SUMMARY: *The town of Abberdoo-Rimey is in shambles. It is smelly and filled with garbage until a green-caped crusader comes to town. Michael Recycle is a recycling superhero who shows the townspeople how to clean up their act and recycle much of their waste.*

Time Needed

This lesson will take several class periods. Suggested scheduling is as follows:

Day 1: **Engage** with How Much Garbage? and **Explore** with Keeping Track of Trash

Day 2: **Explain** with *The Three R's* Read-Aloud and The Three Rs Plus One

Day 3: **Elaborate** with Write a Letter

Day 4: **Evaluate** with *Michael Recycle* Read-Aloud and Three Rs Superheroes

Materials

Per class

- Clear plastic trash bag filled with about 4 lb of clean, safe trash; some of the trash should be reducible (e.g., disposable lunch packaging, disposable plastic water bottles, or paper towels), some should be reusable (e.g., unmarked construction or copy paper), some should be recyclable (e.g., aluminum beverage cans or newspapers), and some should be compostable (e.g., banana peels, apple cores, or leaves).
- Tape

Per student

- 2 pieces of scrap paper
- Disposable gloves (for teacher use)

Student Pages

- The Three Rs Plus One (1 per pair)
- Write a Letter (1 per student)
- My Three Rs Superhero Pledge! (1 per student)

Background

It is easy to throw our garbage in a can, take it out to the curb, and never give it a second thought, but managing solid waste is a growing problem that affects us all. According to the Environmental Protection Agency (EPA), the average American produces over 4 pounds of trash per day, yet only 34% of the solid waste in the United States is recycled (EPA 2011). The rest of the trash goes to a landfill or incinerator, both of which are harmful to the environment. But recycling is becoming easier than ever. There are over 9,000 curbside recycling programs in the United States. It is important to know which materials can be recycled in your area and where to take them. In this lesson, students use the Earth 911 website, which allows you to locate recycling centers in your area by entering the name of an item and your zip code. Recycling is just one of three main ways we can decrease the amount of trash we send to landfills or incinerators. Known as the *three Rs,* these three methods are reduce, reuse, and recycle. *Reduce* means to throw away less, and one of the best ways to do this is to refrain from buying what we really don't need. *Reuse* means to use something again instead of throwing it away. This can be finding a new purpose for an old item or giving the item away so someone else can use it. *Recycle* means to process used materials to make new ones, typically through a recycling center. Food scraps can also be recycled by the process of *composting*. Together, reducing, reusing, and recycling trash can greatly decrease the amount of solid waste that goes to landfills and incinerators. *A Framework for K–12 Science Education* suggests that at an early age students should become aware that some things that we do to live comfortably can affect the world around us, but we can make choices that reduce our impact on the land, air, water, and other living things, such as reducing the amount of trash we throw away by reusing and recycling. This lesson gives students the opportunity to learn how materials we often throw in the trash can be recycled, reused, and composted instead, and how we can reduce the amount of things we use in the first place.

SAFETY

When working with plastic garbage bags filled with trash, make sure the contents are clean, free of sharp objects, and selected by the teacher. Never secure or use trash from unknown sources in activities.

engage

How Much Garbage?

Begin by holding up a clear plastic trash bag filled with about four pounds of trash that might be typical of your classroom, some of which could be recycled, reused, or reduced. Tell students that this is the amount of trash that the average person throws away each day. Ask students to imagine the amount of trash that one person would throw away in a week, a month, a year. That's a lot of garbage!

explore

Keeping Track of Trash

At the beginning of the school day, give each student a piece of scrap paper and ask them to tape it to their desk. Tell them that on this scrap paper, you would like them to draw or write a list of all of the things they throw in the trash that day at school. Remind students several times throughout the day to update their lists. They should include the items they threw away in the classroom, cafeteria, playground, and so on. At the end of the day, take time to discuss their lists. *Ask*

? How much garbage do you think you threw away today? More than four pounds? Less than four pounds?

? How much garbage do you think our whole class threw away today?

Four pounds of trash

- ? Where do you think the classroom garbage goes at the end of each day? Where does it go first? Where does it go next?

Give students time to discuss their ideas with a partner. Ask each pair of students to draw a diagram showing where they think the garbage goes with words and/or pictures, starting with the classroom trash can. Use these diagrams to assess prior knowledge about solid waste disposal.

explain

The Three Rs Read-Aloud

Tell students that they are going to learn more about where the garbage goes after it leaves the classroom and how we might be able to decrease the amount of trash we send there. Show them the cover of the book *The Three R's*, but hide the words "Reduce, Reuse, and Recycle" with sticky notes. *Ask*

- ? Can you name the three Rs? (Some students may already know that the three Rs are reduce, reuse, and recycle.)

 *Determining Importance*

Tell students that as you read pages 2 and 3 you would like them to listen for what each "R word" means. Stop as you encounter each of the R words in the book and ask students to explain what each one means.

SIGNALING WITH THE "R" SIGN

> Connecting to the Common Core
> **Reading: Informational Text**
> CRAFT AND STRUCTURE: K.4, 1.4, 2.4
> KEY IDEAS AND DETAILS: K.2, 1.2, 2.2

As you read pages 4 through 7, *ask* students to listen for where the garbage goes.

- ? Where does the garbage go once it leaves our classroom? (in a dumpster, then in a garbage truck, then to a landfill or incinerator)
- ? How does this compare with the diagram you made?
- ? Does anyone know where the landfill or incinerator is for our trash? (If not, look up the answer together online.)
- ? What is the problem with landfills and incinerators? (They smell bad and are harmful to plants, animals, and people.)

As you read the rest of the book, have students signal (you can have them make the American Sign Language sign for R, which is the middle finger crossed over the index finger) when they hear an example of Paul's family or class reducing, reusing, or recycling. When you read page 23, stop to point out that composting is an example of recycling because the food scraps are made into something new (fertilizer), but composting is different from other kinds of recycling because it is done by natural processes. Note: On page 23, the text refers to fertilizers as "food for plants." This is incorrect because plants produce their own food through the process of photosynthesis. To help prevent any student misconceptions from forming, change the phrase "is food for plants" to "helps plants grow" and the phrase "may become food for plants" to "may become fertilizer for plants."

After reading, show students the four-pound bag of trash you showed them at the beginning of the lesson. *Ask*

? Where would this trash end up if we put it all in the trash can? (a landfill or incinerator)

? What are four ways we could reduce the amount of trash we throw away every day? (reduce, reuse, recycle, and compost)

The Three Rs Plus One

Give each student a copy of The Three Rs Plus One student page. Point out that the headings on the chart indicate four different ways you could deal with the trash in the bag instead of sending it to the landfill or incinerator. You could reduce, reuse, recycle, or compost the items. Wearing disposable gloves, pull each item out of the four-pound trash bag one at a time and discuss what it is. Have students write the name or draw a picture of the item. Then have them discuss the item with a partner or in a small group to determine whether or not it could have been eliminated totally from the trash (e.g., reducing the use of disposable packaging by using an insulated lunch bag and reusable food containers), used again (e.g., reusing paper by writing on both sides), made into something new (e.g., recycling aluminum cans), or turned into valuable soil or fertilizer (e.g., composting yard waste).

Point out that many of the items might have a check mark under more than one heading; for example, a disposable plastic water bottle could be recycled, but it could also be reduced from the trash to begin with by using a sport bottle instead.

You can show students how to use the Earth911 website (*http://earth911.com*) or the iRecycle app (*http://earth911.com/irecycle*) to determine if any of the items are recyclable in your area.

Pulling Items From the Trash

elaborate

Write a Letter

Writing

> Connecting to the Common Core
> **Writing**
> Text Types and Purposes: K.2, 1.2, 2.2
> **Language**
> Vocabulary Acquisition and Use: K.6, 1.6, 2.6

Give each student a copy of the Write a Letter student page. Have them use the chart on The Three Rs Plus One student page to help them write a letter to the person who threw away all of the items in the trash bag to convince them to reduce the amount of trash they throw away. The letter should describe where the trash goes after it leaves the school and what could be done with some items in the bag besides throwing them out. Students should use the terms: *landfill*, *reduce*, *reuse*, *recycle*, and *compost*.

For very young students who are not yet writing, we suggest writing the letter as a class and having the students create a poster with an illustration persuading the person to reduce the amount of trash. For fun, you can have students address their letters to a fictitious name, like Wastey Wally.

Michael Recycle Read-Aloud

Connecting to the Common Core
Reading: Informational Text
Key Ideas and Details: K.2, 1.2, 2.2

Show the students the cover of the book *Michael Recycle*. Have the students look carefully at the illustrations on the cover. *Ask*

? What do you think this book might be about?

? Who do you think Michael Recycle is?

? What does recycle mean? (to make something new out of something old)

Read the book aloud to the class. After reading, *ask*

? What is the lesson the author is trying to teach with this story? (Answers will vary but should include a message about the importance of recycling.)

? Why was Michael Recycle considered a superhero? (He saved the town from too much garbage.)

? How can we be three Rs superheroes? (by reducing, reusing, and recycling at home and at school)

Three Rs Superheroes

Give each student a My Three Rs Superhero Pledge! student page. Have them fill in their name and write or draw several ways they could be a three Rs superhero. (Examples include reducing their use of water bottles, using both sides of paper, and recycling aluminum cans.) Remind them that composting is actually a form of recycling. Then they should draw a picture of themselves as a superhero. For fun, they can come up with their own superhero name, such as The Green Machine, Trash Terminator, or Reuse-It-Man.

Websites

Earth 911
http://earth911.com

Eco-Cycle
http://ecocycle.org/ecofacts

More Books to Read

Davis, D., and J. Peck. 2011. *The green mother goose: Saving the world one rhyme at a time.* New York: Sterling.
Summary: In this playful picture book, Old Mother Hubbard shops with cloth grocery bags; Old King Coal works to keep our skies smoke free; and Hickety Pickety is now a cage-free hen. Illustrated with eco-friendly collages created from ticket stubs, newspapers, and other reused items.

Gibbons, G. 1992. *Recycle! A handbook for kids.* New York: Little, Brown.
This book explores the problems we face as our landfills become more and more full. The author explains that many of the materials we use are recyclable and can be used to make new products.

Green, J. 2005. *Why should I recycle?* Hauppauge, NY: Barron's Educational Series.
Summary: In this fictional narrative a teacher shows his students the importance of recycling by separating his own trash. This book will help teach children the importance of preserving our resources and the reasons why we should recycle.

Inquiry Place

Have students brainstorm questions about the three Rs. Examples of such questions include

? How can you turn used school supplies into useful new items? Try it!

? What products cannot be recycled at your local recycling center? Why not? Research it!

? What items are not allowed in your landfill? Where can you dispose of them instead? Research it!

? What new products are made from used plastic bottles, aluminum cans, plastic bags? Research it!

Have students select a question to research as a class, or have groups of students vote on the question they want to research as a team. Students can present their findings at a poster session or gallery walk.

Inches, A. 2008. *I can save the Earth! One little monster learns to reduce, reuse, and recycle.* New York: Little Simon.
Summary: In this book a little green monster learns how to go green by using the three Rs: reduce, reuse, and recycle.

Inches, A. 2009. *The adventures of a plastic bottle: A story about recycling.* New York: Little Simon.
Summary: This story takes young readers through the experiences of a plastic bottle as it goes through its life at the refinery, sitting on the store shelf, lying in the trash, and finally being reborn at the recycling plant.

Wallace, N. 2003. *Recycle every day!* Tarrytown, NY: Marshall Cavendish.
Summary: Minna is an environmentally conscious bunny who sets out to create a community recycling program. Through Minna's efforts children can learn the importance and impact recycling can have on our environment.

Reference

Environmental Protection Agency. 2011. *Municipal solid waste generation, recycling, and disposal in the United States: Facts and figures for 2010. www.epa.gov/osw/nonhaz/municipal/pubs/msw_2010_rev_factsheet.pdf*

Name: ____________________

The Three Rs Plus One

Directions: List or draw each item in the bag of trash. Could you reduce, reuse, recycle, or compost it? Place a check mark in the correct column(s) below.

Item	Reduce	Reuse	Recycle	Compost

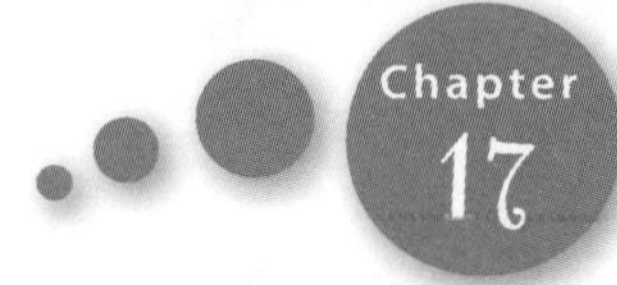

Write a Letter

Write a letter to the person who threw away all of the items in the trash bag to convince them to reduce the amount of trash they throw away. In your letter, be sure to

- Tell them where the trash goes after it leaves your school,
- Describe what they could do with some of the items in the bag besides throwing them in the trash, and
- Use the terms *landfill, reduce, reuse, recycle,* and *compost*

Dear ______________________________,

____

____

____

____

____

____

____

____

____

Your Friend,

____

My Three Rs Superhero Pledge!

I, ______________________ pledge to be a Three Rs Superhero by:

____

Me as a Three Rs Superhero

What Will the Weather Be?

Description

The weather is something that affects us every day. It can determine what we wear, where we go, and what we do. In this lesson, students explore various weather instruments and learn how they help meteorologists predict the weather.

Suggested Grade Levels: 3–5

LESSON OBJECTIVES *Connecting to the Framework*

EARTH AND SPACE SCIENCES

CORE IDEA ESS2: EARTH'S SYSTEMS

ESS2.D: WEATHER AND CLIMATE

By the end of grade 5: Weather is the minute-by-minute to day-by-day variation of the atmosphere's condition on a local scale. Scientists record the patterns of the weather across different times and areas so that they can make predictions about what kind of weather might happen next.

Featured Picture Books

TITLE: *Come On, Rain!*
AUTHOR: **Karen Hesse**
ILLUSTRATOR: **Jon J. Muth**
PUBLISHER: **Scholastic Press**
YEAR: **1999**
GENRE: **Story**
SUMMARY: *A young girl awaits the rain during a heat wave in the city.*

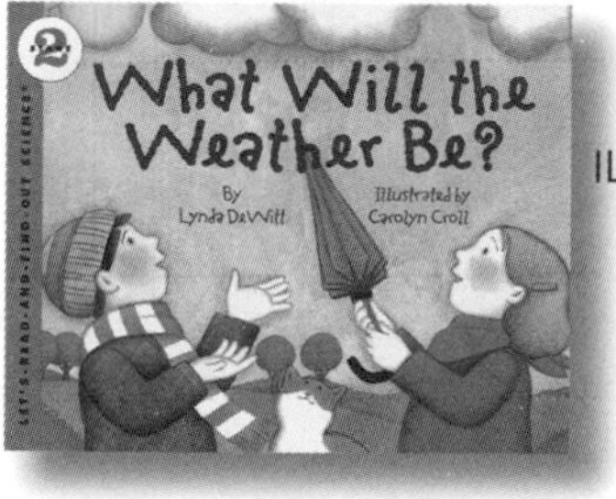

TITLE: *What Will the Weather Be?*
AUTHOR: **Lynda DeWitt**
ILLUSTRATOR: **Carolyn Croll**
PUBLISHER: **HarperCollins**
YEAR: **1993**
GENRE: **Non-Narrative Information**
SUMMARY: *This book explains the basic characteristics of weather—temperature, humidity, wind speed and direction, air pressure—and how meteorologists gather data for their forecasts.*

Time Needed

This lesson will take several class periods. Suggested scheduling is as follows:

Day 1: **Engage** with *Come on Rain!* Read-Aloud and **Explore** with Weather Instruments

Day 2: **Explain** with *What Will the Weather Be?* Read-Aloud

Day 3: **Elaborate** with Weather Instrument Placement and Daily Reports

Day 4 and beyond: **Evaluate** with Weather Report Skits

Materials

- Alcohol thermometer
- Wind vane
- Anemometer
- Hygrometer
- Barometer
- Daily Reports sheet to project (for teacher use)
- (Optional) Video recording device

Student Pages

- Weather Instruments
- Daily Reports
- 4-3-2-1 Weather Report Skit Rubric

Background

A Framework for K–12 Science Education suggests that by the end of grade 5 students have some basic understandings about weather, including the fact that scientists record the weather across different times and areas to predict what might happen next. Even with all of the advances in technology and instrumentation, it is impossible to be 100% accurate with the weather forecast all the time. The intricacies of weather are vast and can be difficult to track. However, knowing some basic information about weather and how it works, can give you an idea of what to expect the weather to do.

Most changes in weather occur along *fronts*, which are the boundaries between two different air masses. A *cold front* occurs where cold air pushes against warm air. These cold fronts usually quickly move the warm air up and out of the way. The rising air carries water vapor up and, as it rises, it turns into liquid water droplets that clump together and form clouds. The clouds grow big and dark and then it rains, or snows if it is cold enough. Cold fronts typically cause sudden storms that do not last long. After a cold front passes, the sky usually clears and the weather is colder. *Warm fronts* occur where warm air pushes against cold air. Warm fronts typically move more slowly than cold fronts and therefore change the weather gradually. They cause mild weather, such as a light drizzle. After a warm front passes, the sky usually clears and the weather is warmer.

Meteorologists, scientists that study and predict the weather, use a variety of instruments to help them predict where these fronts will form. They measure the temperature of the air using *thermometers* and record the temperature in degrees Fahrenheit or degrees Celsius. They find out the direction the wind is blowing (north, south, east, or west) by using *wind vanes,* and they measure wind speed (in miles per hour or kilometers per hour) using *anemometers.* A *hygrometer* tells them the humidity, or how much water vapor is in the air, which is expressed as a percentage. A *barometer* measures the air pressure in millimeters of mercury (mmHg). These measurements are taken by meteorologists all over the world to help them predict where fronts will form and what our weather will be.

engage

Come on Rain! Read-Aloud

> Connecting to the Common Core
> **Reading: Literature**
> Craft and Structure: 3.4, 4.4, 5.4
> Integration of Knowledge and Ideas: 3.7, 4.7, 5.7

Visualizing

Show students the cover of the book *Come On, Rain!* and introduce the author and illustrator. Tell students that as you read the first few pages, you would like them to imagine what the narrator is experiencing—what might it feel like, smell like, taste like, sound like, and look like where she is? Read the book aloud, stopping at page 11 to *ask*

? What do you think the girl in the story is feeling on these first few pages? (Answers will vary but may include hot, sweaty, drained, bored, and tired of the heat.)

? What does it smell like there? (like hot tar and garbage, p. 10)

? What does it sound like there? (quiet except when a truck goes by, pp. 6 and 7).

? What words does the author use to help the reader imagine how hot the setting is in the story? (Answers will vary, but here are a few to point out to students: endless heat, p. 4; parched, p. 5; sizzling like a hot potato, p. 6; crackling dry path, p. 8; stuffy cave, p. 9; slick with sweat and senseless in the sizzling heat, p. 11)

Model how to use the context of the sentence and the story to figure out what unknown words mean.

? How do the illustrations help establish the setting? (Answers will vary but may include the fact that you can tell by the illustrations that they are in a city, the color yellow is a "warm" color that helps create the mood of sizzling heat, and the plants are droopy.)

Inferring

Read pages 12–25 and *ask*

? What do you think the Mammas are going to do? Turn and talk.

Read pages 26–27 to find out what the Mammas do in the story.

Visualizing: Sketch to Stretch

Before reading pages 28 and 29, tell students that you are not going to show them the illustration on the next page yet because you want them to close their eyes and imagine what the scene looks like as you read the author's words. Read aloud pages 28 and 29, then ask students to make a sketch of what the scene looks like in their imagination. Students can share their sketches with each other, and then you can reveal the illustration.

Making Connections

Finish reading the book aloud, then *ask*

? Can you think of a time that you have hoped and wished for rain like Tessie, the girl in the story? (Have students share with a partner.)

? Have you ever played or danced in the rain? (Have students share with a partner.)

? What clues did Tess have that the rain was coming in the story? (gray clouds moving in, the wind blows "bolder")

? Have you ever had an experience where you could sense that the weather was going to change very quickly? What were the clues? (Have students share their experiences with a partner.)

explore

Weather Instruments

Tell students that sometimes we are able to tell if the weather is going to change, just by being outside right before it happens. But the most reliable way to predict changes in weather is through measuring the weather conditions. Tell students that meteorologists are men and women with specialized education who use scientific principles to explain, understand, observe, or forecast weather and that they use measurements from various tools to predict the weather days ahead of time.

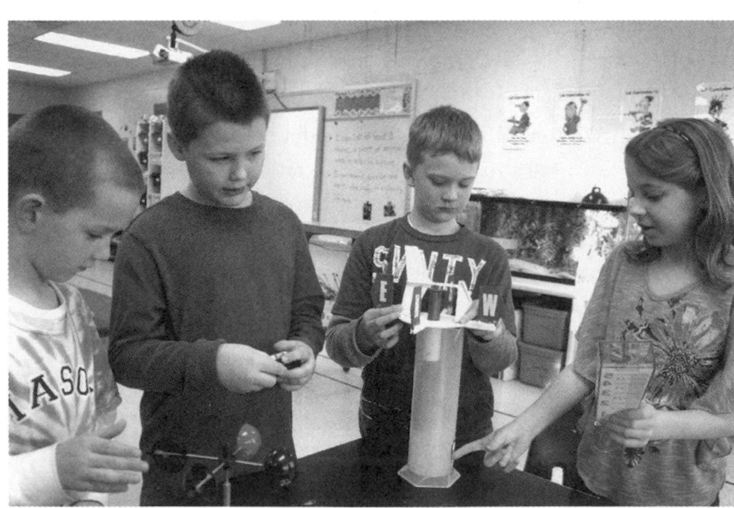

EXPLORING WEATHER INSTRUMENTS

SAFETY

Use caution when working with sharp items as they can cut or puncture skin (e.g., glass instruments).

In advance, set up five stations, each containing one of the following weather instruments: alcohol thermometer, wind vane, anemometer, hygrometer, and barometer. Give each student a copy of the Weather Instruments student page. Assign students to work in groups of four to five. Explain that with their group they will be looking at each instrument and trying to figure out what it measures. Caution students to be careful with the weather instruments because most of them are fragile. Allow groups to move from station to station, completing the "What I Think It Measures" column on the Weather Instruments student page.

explain

What Will the Weather Be? Read-Aloud

 ***Determining Importance***

Connecting to the Common Core
Reading: Informational Text
CRAFT AND STRUCTURE: 3.4, 4.4, 5.4
KEY IDEAS AND DETAILS: 3.1, 4.1, 5.1

Tell students that you have a book that can help them figure out what all of these weather instruments measure and how they help predict the weather. Show students the cover of *What Will the Weather Be?* Tell students that as you read, you would like them to signal (e.g., raise their hands) when they see or hear one of the weather instruments. Pause after you come to each one in the text and discuss what it measures, and have students fill out the "What It Measures" column of the Weather Instruments student page. The correct answers are as follows:

- Thermometer: temperature
- Wind vane: wind direction
- Anemometer: wind speed
- Hygrometer: humidity (amount of water vapor in the air)
- Barometer: air pressure

Next, reread pages 9–18 of *What Will the Weather Be?* and tell students you would like them to listen for the answers to the following questions:

Page 11: What causes changes in weather? (fronts, where new air pushes against old air)

Page 12: What kind of weather is the result of a cold front, where cold air pushes against warm air? (sudden heavy rain and storms that don't last long)

Page 16: What kind of weather is the result of a warm front where warm air pushes against cold air? (gradual change to light rain or drizzle, slowly changing into clear skies)

Pages 18–29: How does taking measurements with these instruments help meteorologists predict changes in weather? (Meteorologists try to predict where fronts will form because that is where most changes in weather occur. Measuring temperature, air pressure, wind direction and speed, humidity, and air pressure miles away helps meteorologists predict where fronts will form and therefore predict the weather.)

Making Connections: Text-to-Text

Ask students to think back to the story of *Come on, Rain!* Ask

? What kind of front made the weather change in Tessie's neighborhood? (Students should be able to determine it was a cold front because the weather changed very quickly, the rain was heavy, and it didn't last long.)

elaborate

Weather Instrument Placement

Tell students that you want them to set up these instruments to take actual measurements of your local weather conditions. In their group, ask them to discuss where on the school grounds each instrument should be placed to get the most accurate measurements; for example, indoors or outdoors, in sunlight or shade, near the building or away from the building, up high or down low. Discuss groups' ideas as a whole class and come to a consensus on a location for each instrument. Then place the instrument in the designated locations.

Daily Reports

Tell students that you would like them to practice recording the measurements of each of the weather instruments where they placed them. Give each student a Daily Reports student page. Record the date, time, and current weather conditions together (e.g., rainy, windy, clear, stormy, cloudy). Visit each instrument together and demonstrate for students how to read each one and record the measurements with the proper units. Have each student record the measurements on their own Daily Reports student page. Then show students how to access these same measurements taken by the National Weather Service (NWS) by entering the school zip code on the home page of the NWS website (*http://weather.gov*). Have them record those measurements in the "NWS" column of the student page and compare them with the measurements they took. If there are big discrepancies between their measurements and those on the NWS website, they may need to try to read the weather instrument again.

For the next week or so, assign each group of students an instrument to check each day until all groups have taken measurements with each instrument. They should use their Daily Reports student pages to record the date, time, measure-

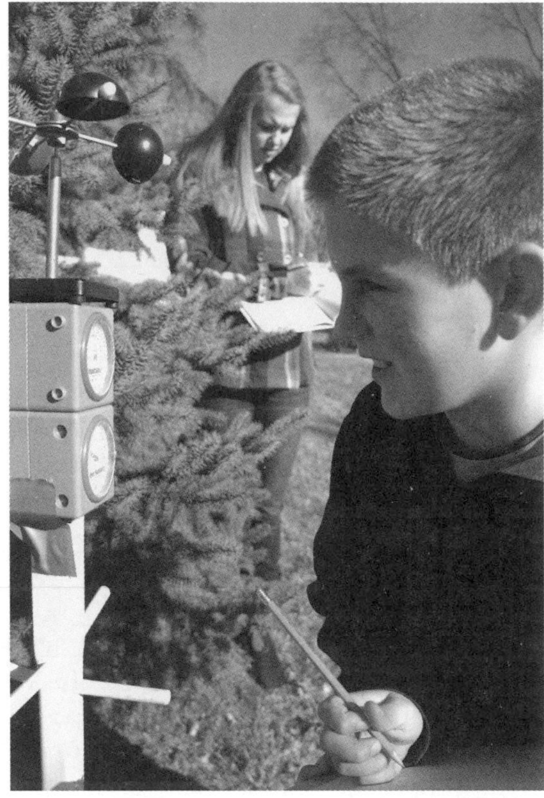
TAKING READINGS

ments, and weather conditions each day. They should also record the measurements posted on the NWS website. Project an image of the Daily Reports student page and have each group add their measurement, so everyone can fill in their own data table for each day. Discuss the weather conditions as they relate to the measurements on a daily basis. Also discuss the forecast given on the NWS website and how these measurements help meteorologists predict what kind of weather we can expect the next day.

evaluate

Weather Report Skits

Connecting to the Common Core
Speaking and Listening
PRESENTATION OF KNOWLEDGE AND IDEAS: 3.4, 4.4, 5.4

Ahead of time, download video of some daily weather forecasts from your local television stations' websites to show your students. Tell students that their teams are going to create a weather report skit that incorporates all of the measurements taken from the weather instruments they have placed around the school as well as a weather forecast based on information they can find on the NWS website. Tell them that as they watch the meteorologists, they should listen for how they report the various measurements of current weather conditions, what tools they use (maps, radar, video of weather events, and so on), and how they effectively communicate the weather conditions and forecast to their viewers. Also note the length of the forecast. Meteorologists usually have a tight time constraint for reporting the weather conditions and forecast. Show a weather forecast and discuss, then have teams create their skits. You may want to set a time limit of five minutes or less for your students' skits. Students can act the skits out live or videotape them and play them for the class. Give each team a copy of the 4-3-2-1 Weather Report Skit Rubric beforehand so they know how you will be evaluating the skits.

Websites

National Weather Service
http://weather.gov

National Oceanic and Atmospheric Administration (NOAA) Links for Kids
www.nws.noaa.gov/pa/forkids.php

NOAA weather terms
www.erh.noaa.gov/box/glossary.htm

More Books to Read

Sherman, J. 2003. *Flakes and flurries: A book about snow*. Mankato, MN: Picture Window Books.
Summary: You can make a snowball. You can make a snowman. But do you know what makes snow? Solve the puzzle by reading all about snow.

Thomas, R. 2005. *Rumble, boom! A book about thunderstorms*. Mankato, MN: Picture Window Books.
Summary: This is no regular rainstorm. After the lightning, sounds rumble and boom. It's a thunderstorm! Learn more about the giant flashes of light

Inquiry Place

Have students brainstorm questions about weather. Examples of such questions include

- ? How accurate is your local forecast? Investigate it!
- ? What causes lightning? Research it!
- ? How do tornadoes form? Research it!
- ? What causes thunder? Research it!

Then have students select a question to investigate or research as a class, or have groups of students vote on the question they want to investigate or research as a team. Have students present this information on a poster. Students will then share their findings in a poster session or gallery walk.

and what makes thunder crash in this book about a building thunderstorm.

Thomas, R. 2004. *Sizzle! A book about heat waves.* Mankato, MN: Picture Window Books.
Summary: Wow, it's hot! No rain in sight, and the temperatures are soaring. Find out what happens in the city and in the country when a heat wave hits.

Name: ______________________ *Date:* __________________________

Weather Instruments

Instrument	What I Think It Measures	What It Measures
Thermometer		
Wind Vane		
Anemometer		
Hygrometer		
Barometer		

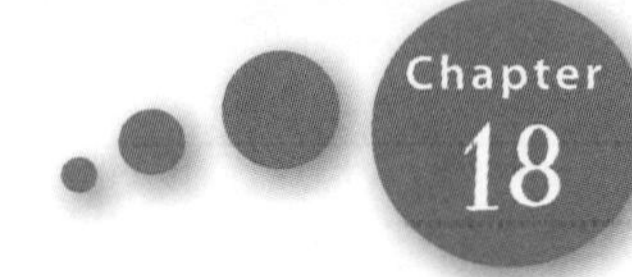

Name: ______________________________

Daily Reports

Daily Reports	Date: ____ Time: ____ Conditions: ____		Date: ____ Time: ____ Conditions: ____		Date: ____ Time: ____ Conditions: ____		Date: ____ Time: ____ Conditions: ____		Date: ____ Time: ____ Conditions: ____	
Instrument	**Class**	**NWS**	**Class**	**NWS**	**Class**	**NWS**	**Class**	**NWS**	**Class**	**NWS**
Thermometer										
Wind Vane										
Anemometer										
Hygrometer										
Barometer										

4-3-2-1 Weather Report Skit Rubric

Names: ____

____

With your team, create a weather report skit using the data from your weather instruments as well as information from the National Weather Service website. Your skit should include:

4 Points: A report on the current weather conditions using readings from your class's weather instruments—including temperature, wind speed and direction, humidity, and barometric pressure

4 3 2 1 0

3 Points: A forecast of the local weather conditions for the next three days using information from the National Weather Service website

3 2 1 0

2 Points: A written script of the skit

2 1 0

1 Point: A tip for helping viewers prepare for the coming weather

1 0

Score:_______ /10

Sunsets and Shadows

Description

In this lesson, students make observations of sunrise and sunset and learn that they are caused by Earth's rotation. They also learn about the effect of Earth's rotation on length and direction of shadows as well as the illusion that the Sun and stars appear to move across the sky.

Suggested Grade Levels: 3–5

LESSON OBJECTIVES *Connecting to the Framework*

EARTH AND SPACE SCIENCES

Core Idea ESS1: Earth's Place in the Universe

ESS1.B: Earth and the Solar System

By the end of grade 5: The orbits of Earth around the Sun and of the Moon around Earth, together with the rotation of Earth about an axis between its North and South poles, cause observable patterns. These include day and night; daily and seasonal changes in the length and direction of shadows; phases of the Moon; and different positions of the Sun, Moon, and stars at different times of the day, month, and year.

Featured Picture Books 

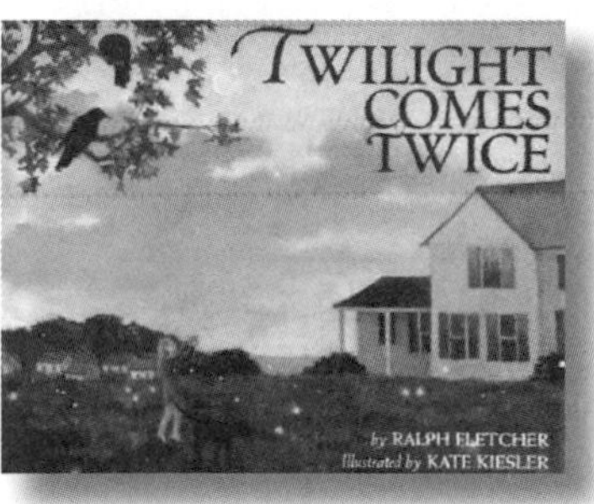

TITLE: ***Twilight Comes Twice***
AUTHOR: **Ralph Fletcher**
ILLUSTRATOR: **Kate Kiesler**
PUBLISHER: **Clarion Books**
YEAR: **1997**
GENRE: **Story**
SUMMARY: *This book combines poetic language with beautiful paintings to show the happenings at dusk and dawn.*

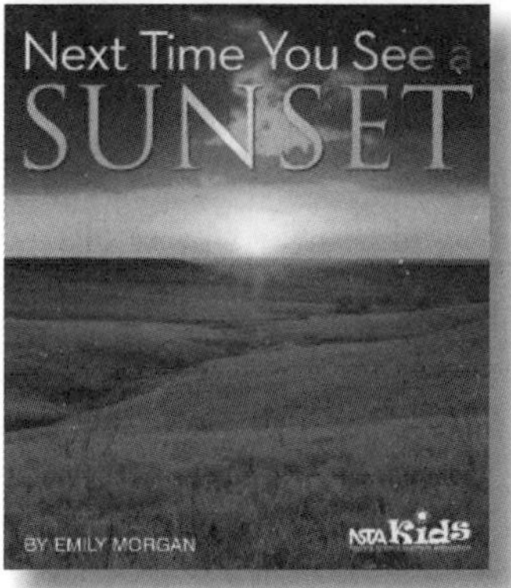

TITLE: ***Next Time You See a Sunset***
AUTHOR: **Emily Morgan**
PUBLISHER: **NSTA Press**
YEAR: **2013**
GENRE: **Non-Narrative Information**
SUMMARY: *This book from the* Next Time You See *series provides explanations of what is really happening at sunset and encourages a sense of wonder about this daily phenomenon.*

Time Needed

This lesson will take about a week. Suggested scheduling is as follows:

Day 1: **Engage** with *Twilight Comes Twice* Read-Aloud and Sunrise/Sunset Take-Home Page

Day 2: **Explore** with Changing Shadows Part 1

Day 3: **Explain** with *Next Time You See a Sunset* Read-Aloud and Changing Shadows Part 2

Day 4: **Elaborate** with Modeling Earth's Rotation

Day 5: **Evaluate** with Sunsets and Shadows Assessment

Materials

For* Twilight Comes Twice *Read-Aloud (per student)

- 3"× 5" or 4" × 6" index cards

For Changing Shadows Part 1

- Sidewalk chalk (per pair)
- Clock or watch (for teacher use)

For Modeling Earth's Rotation

Per class

- Lamp that shines in all directions

Per team of three to four students

- Globe
- Small sticky notes
- Small piece of modeling clay

Student Pages

- Sunrise/Sunset Take-Home Page
- Changing Shadows Part 1
- Changing Shadows Part 2
- Sunsets and Shadows Assessment

Background

A Framework for K–12 Science Education suggests that by the end of grade 5 students should understand that because of the position and motion of the Earth, Sun, and Moon we experience regular, predictable patterns including day and night, Moon phases, and seasons. This lesson focuses on the phenomena we observe as a result of the rotation of the Earth: day and night, changing position of the Sun in the sky, and changes in length and direction of shadows. It is important for students to understand that even though the Sun appears to move across the sky in a daily cycle and the stars appear to move across the sky at night, it is Earth's rotation that causes this illusion. The Sun seems to "rise" in the sky as your place on Earth turns toward the Sun, reaches its maximum height at midday, and "sets" as your place on Earth then turns away from the Sun. While it is day in the Western Hemisphere, it is night in the Eastern Hemisphere and vice versa. The Sun is always shining. Students are introduced to this concept by focusing on the time between day and night, twilight. They watch a sunrise or sunset for homework and then use globes and lamps to find the line between day and night, where people who live in that location on Earth would be experiencing twilight. Another way to sense Earth's rotation is to observe the change in the length and direction of shadows throughout the day.

engage

Twilight Comes Twice Read-Aloud

Connecting to the Common Core
Reading: Informational Text
KEY IDEAS AND DETAILS: 3.1, 4.1, 5.1

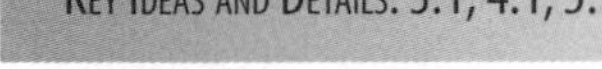

Questioning

Show students the cover of *Twilight Comes Twice* and introduce the author and illustrator. Read the title and *ask*

? What is twilight?

? What do you think the title of the book means? Tell students that as you read the book aloud, you would like them to listen for what the word *twilight* means and how it comes twice. Then read the book aloud to students.

After reading, *ask*

? What is twilight? (the time between day and night)

? Does it really come twice? (Yes, it happens in the morning and the evening.)

? What is morning twilight called? (dawn)

? What is evening twilight called? (dusk)

? Does everyone experience dusk and dawn at the same time? (Answers will vary.)

Then *ask*

? What causes dawn and dusk to occur twice each day? What is actually happening at those times? (Answers will vary.)

Give each student an index card and have them explain what causes dawn and dusk, using both words and pictures. Collect the cards and use them as a preassessment.

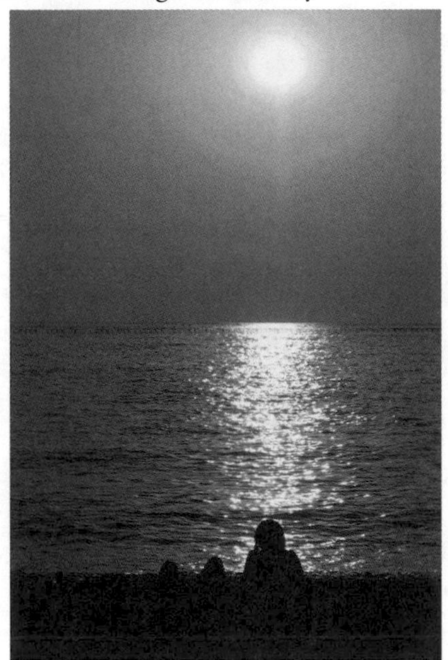

WATCHING A SUNSET

Making Connections: Text-to-Self

Ask

? Have you ever watched a sunrise or sunset?

? Where were you?

? What did it look like?

SAFETY

- Bring some form of communication, such as a cell phone or two-way radio, when taking students outdoors, in case of emergency.
- Warn students to never look at the Sun directly or through reflected light from it. Direct sunlight can seriously injure eye tissue.

Give students the Sunrise/Sunset Take-Home Page. Tell them that their homework is to go outside with an adult helper sometime in the next few days and observe a sunrise or sunset. (You may want to provide a resource for sunrise/sunset times, such as the printable calendars available at *www.sunrisesunset.com*). Students will need to either take a photograph or draw a picture of it and then write down some of their observations and questions. The student page includes a brief note to parents, a list of items to bring, tips for observing a sunrise or sunset, and a chart for observations and wonderings chart.

After students have completed their Sunrise/Sunset Take-Home Page, have them share their pictures and some of their observations and wonderings in small groups. Invite students to share their most compelling wonderings with the rest of the class. Then display the pictures in the classroom or hallway.

explore

Changing Shadows: Part 1

This phase of the lesson is about shadows and includes observations that take place over the course of a day. Be sure to do the following activities on a clear, sunny day. If possible, take a photograph of one of the student's shadows in the morning, midday, and afternoon as an example to refer to the next day.

Morning: Tell students that they are going to learn some things about the relationship between Earth and the Sun by making some observations of their shadows. First thing in the morning, give each pair of students a piece of sidewalk chalk and take them outdoors to a place where they can trace their shadows, such as the playground or sidewalk. Be sure to choose a place that will remain untouched throughout the day so the tracings will not be erased. Have students spread out so their shadows do not overlap. As one partner stands very still, have the other partner trace his or her shadow. After tracing, have them write the name of the person on the feet of the shadow and the time (including "a.m." or "p.m.") in the head of the shadow. Next have the partners switch and repeat the same procedure. Cautioning students to not look directly at the Sun, have them point in the direction of the Sun in the sky and notice that their shadows are in the opposite direction of the Sun.

Midday: As close to noon as possible, go outside to the same place where they traced their shadows in the morning and have students repeat the procedure with their feet in the same exact spot they were in the morning. Have students compare the size and direction of the morning shadows with the midday shadows. Students should notice that the midday shadows are shorter than the morning shadows and that the two shadows are in different directions. Have students point in the direction of the Sun in the sky and notice that their shadows are, once again, opposite the direction of the Sun.

Afternoon: As late as possible in the afternoon, take your students back to the place where they traced their shadows and have them repeat the procedure again. Have students compare the size and direction of the three shadows and notice the position of the Sun in the sky. Give each student a copy of the Changing Shadows Part 1 student page and have them answer the four questions on the page while they are outdoors observing their traced shadows.

1. When was your shadow the longest? (Students should record the morning or afternoon time when their shadows were traced, depending on which one was longest.)
2. When was it the shortest? (Students should record the midday time when their shadows were traced.)
3. How did the direction of your shadow change throughout the day? (Students should recognize that the shadows were in different directions. At this point, they don't necessarily need to use the words *east* or *west*. The important thing is that they notice that the shadows in the morning and afternoon were in different

TRACING SHADOWS

directions and the midday shadow was somewhere in between.)

4. How did the position of the Sun in the sky change throughout the day? (Students should note that the Sun appeared lower in the sky in the afternoon and morning and higher in the sky at midday. Although they may not use the words *east* and *west* yet, they should describe what part of the sky the Sun was in at different times using landmarks such as the school, road, playground, and so forth.)
5. Why do you think your shadow changed throughout the day? (Answers will vary but should include the idea that as the position of the Sun in the sky changed, the direction and size of their shadows changed.)

With partners or in small groups, have students discuss their ideas about why their shadows changed throughout the day. Tell students that the next day you will be sharing a book with them that will help them understand more about this phenomenon.

explain

Next Time You See a Sunset Read-Aloud

> Connecting to the Common Core
> **Reading: Informational Text**
> KEY IDEAS AND DETAILS: 3.1, 4.1, 5.1

Show students the cover of *Next Time You See a Sunset.* Introduce the author, Emily Morgan, and tell students that she has always been fascinated by the time between day and night. Tell them that this is a nonfiction book that can help them learn more about the reasons for dawn and dusk as well as the answers to some of the questions they have about sunrises, sunsets, and changing shadows. Read the book aloud, stopping after reading page 17.

Questioning

After reading page 17, *ask*

- ? What is really happening during a sunset? (Your place on Earth is turning away from our star, the Sun.)
- ? What is really happening during a sunrise? (Your place on Earth is turning toward our star, the Sun.)

Read pages 18 and 19, which explains the changing position of the Sun in the sky throughout the day. *Ask*

- ? How does this explanation relate to the position of the Sun in the sky yesterday when we traced our shadows? (Students should connect this explanation to their observation of the changing position of the Sun in the sky and should now be able to figure out which direction is east and which is west based on their experience with the Sun and shadows.)

Continue reading and stop after you read page 23, which explains the changes in the size and direction of shadows throughout the day. *Ask*

- ? When were your shadows the shortest? (in the middle of the day, around noon)
- ? Where was the Sun at that time? (right above us)
- ? When were your shadows the longest? (in the morning and afternoon)
- ? Where was the Sun at those times? (in the eastern sky in the morning and in the western sky in the afternoon)
- ? What causes this change in position of the Sun in the sky? Is the Sun really moving across the sky? (No, it just appears that way because Earth is rotating.)
- ? Do you think your shadows would look similar if you did this again tomorrow? (Yes, because Earth is always turning in the same direction.)

Next *ask*

- ? We know the sun always rises in the eastern sky and sets in the western sky, but in which direction do the moon and stars rise and set?

Allow student time to think about this and share their ideas with partners. Then read page 25, which explains that the stars "appear" to move across the sky from east to west. Explain that from Earth everything in the sky (Moon, stars, planets) "rises" in the east and "sets" in the west because Earth is always turning in the same direction.

Changing Shadows: Part 2

Finish reading the book aloud. Then give students the Changing Shadows Part 2 student page. Have them answer the questions using their own experience and what they learned from the book. Students should be able to explain that the two people are not running at the same time of day because their shadows are very different sizes. They should infer that Person A is running in the morning or afternoon because his shadow is very long. The sun appears lower in the sky at those times of day, creating longer shadows. They should infer that Person B is running more toward the middle of the day, when shadows are shorter. The Sun appears higher in the sky at those times of day, which makes shadows shorter. All of this occurs because of Earth's rotation.

NOTE: It is important to know that size and direction of shadows also change seasonally. But because the objective of this lesson is for students to learn the effects and patterns caused by Earth's rotation, it is not necessary to introduce the seasonal changes in shadows due to Earth's tilt and revolution at this time.

elaborate

Modeling Earth's Rotation

SAFETY

- When working with electrical devices, lightbulbs, etc., keep away from water or other liquids to prevent electric shock.
- Use caution when working in a darkened area. Make sure all trip/fall hazards are removed.
- Be careful working with lamp bulbs. Skin can be burned from the high temperatures and heat.

These modeling activities work best if your classroom is as dark as possible, with the lamp the only light source in the room. Close blinds, pull down the window shades, and cover windows with black paper to make the room as dark as you can.

Provide each team of three to four students with a globe. Place a lamp that shines in all directions in the center of the room and have students form a circle around it with their globes. Tell them that the globes represent the Earth and the lamp represents the Sun. *Ask*

? Does the Sun ever stop shining? (no)

Tell them that because the Sun never stops shining, you will keep the lamp lit in the center of the room. Next ask students to locate the areas of land and water on the globe and then pinpoint their location using a small sticky note. *Ask*

? Can you model daytime in your location? (They should make their location on Earth face the lamp.)

? What are some places that are experiencing night as we are experiencing day? (places directly opposite)

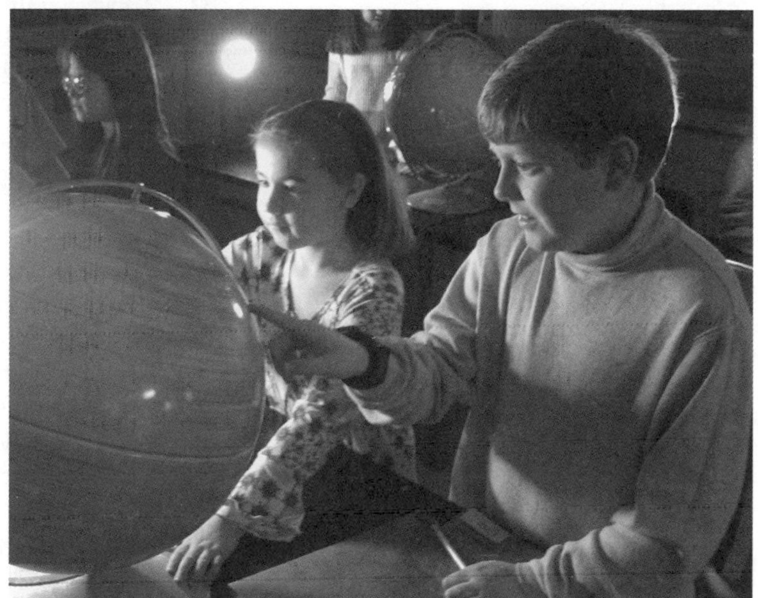
MODELING EARTH'S ROTATION

? Can you find the arrow on the globe that points in the direction that Earth turns? Slowly spin the globe in that direction. (Students should be spinning the globe counterclockwise.)

Explain that because the Earth is always turning in the same direction, we always see the Sun in the eastern sky in the morning (as our location turns toward the Sun) and in the western sky in the evening (as our location turns away from the Sun).

Have students use small sticky notes to label the parts of the Earth experiencing day (facing the lamp), night (turned away from the lamp), sunrise (the line between light and darkness on one side that would be turning toward the lamp), and sunset (the line between dark and light that would be turning away from the lamp). Check to see that students' globes are properly labeled.

? Provide each team of students with a small piece of clay to place on their globe representing a person standing on Earth. Have them practice using the globe, lamp, and clay to model the answers to the following questions:

- **?** Why does a person's shadow look different throughout the day? (Light from the Sun is coming at the person in different directions throughout the day because the Earth is rotating.)
- **?** When is it the longest, shortest? (Shadows are longest in the morning and afternoon and shortest in the middle of the day.)
- **?** Can you model sunrise for a clay person standing on Earth? (Check students' models to make sure the clay person is just entering the lamplight as the globe is rotating counterclockwise.)
- **?** Can you model sunset for the clay person standing on Earth? (Check students' models to make sure the clay person is just leaving the lamplight as the globe is rotating counterclockwise.)

evaluate

Sunsets and Shadows Assessment

Writing

> Connecting to the Common Core
> **Writing**
> Text Types and Purposes: 3.2, 4.2, 5.2

Give each student a copy of the Sunset and Shadows Assessment, which includes two questions about sunrise/sunset and shadows:

- **?** Why is it not scientifically accurate to say the Sun is "going down" at sunset? (Students should explain that the Sun is not moving up and down in the sky. When sunset happens, that place on Earth is turning away from the Sun into the darkness of space, which makes the sun *appear* to go down.)
- **?** If you go outside on a sunny day, you will see your shadow. Sometimes your shadow is longer than you are, and other times it is shorter than you are. How can this difference in the length of your shadow be explained? Use a labeled drawing to help explain your answer. (Students should explain that the length of their shadow changes throughout the day as the position of the Sun in the sky changes and that the change in the position of the Sun in the sky is due to Earth's rotation. Their drawings should support this explanation.)

Websites

Day and Night World Map
www.timeanddate.com/worldclock/sunearth.html

Sunrise and Sunset Times for Your Location
www.sunrisesunset.com

U.S. Naval Observatory: Day and Night Across the Earth
http://aa.usno.navy.mil/data/docs/earthview.php

More Books to Read

Bailey, J. 2006. *Sun up, Sun down: The story of day and night*. Minneapolis, MN: Picture Window Books.
Summary: This book follows the path of the Sun from dawn to dusk. Cartoonish illustrations and entertaining text make the concept of day and night fun for the reader.

Branley, F. 1986 *What makes day and night*. New York: Harper & Row.
Summary: This *Let's-Read-and-Find-Out Science* book explains what causes day and night and includes instructions for demonstrating this phenomenon.

Schuett, S. 1997. *Somewhere in the world right now*. New York: Dragonfly Books.
Summary: This book takes children around the world to show what's going on at the exact same moment in other areas of the world. A time zone map on the endpapers, which includes the times and names of places shown in the pictures, allows readers to follow the action around the globe.

Inquiry Place

Have students brainstorm questions about day and night. Examples of such questions include

- ? How does a sundial work? Research it, then try it!
- ? Does sunrise/sunset always happen at the same time every day? Investigate it, or research it!
- ? What is daylight savings time, and why do some places on Earth use it? Research it!

Then have students select a question to research or investigate as a class, or have groups of students vote on the question they want to research or investigate as a team. Students can present their findings in a poster session or a gallery walk.

Sunrise/Sunset Take-Home Page

Dear Parent,

At school, we are studying Earth's patterns and cycles. Your child's homework assignment is to go outside with an adult helper and observe a sunset or a sunrise. As you look at the sky with your child, help him or her record "Observations" and "Wonderings" (questions) on the attached chart. Also, have your child either take a photograph or draw a picture of the sunrise or sunset. Below is a list of items to bring outside with you and some tips on making the most of this experience. The purpose of this assignment is to give your child the opportunity to observe this daily phenomenon firsthand and wonder about it. In class, we will be building on this experience by reading about it and modeling the Earth and Sun relationship.

This assignment is due by_______________ .

Items to bring:

- Sunrise/Sunset Take-Home Page
- Flashlight
- Clipboard or notebook
- Camera or art supplies
- Pen or pencil

Sunrise/sunset viewing tips:

1. Do not look directly at the Sun. Sunlight can cause eye damage.
2. Find a place, without a lot of trees or buildings, where there is a clear view of the eastern sky (sunrise) or western sky (sunset). Be sure to bring a flashlight with you so you can find your way before sunrise or after sunset.
3. Watch the colors of the sky change. Discuss what colors you see, how the temperature of the air feels, and how your shadows look. Share your ideas and wonderings about what is happening.

Observe a sunrise or sunset with an adult helper. Record your observations, wonderings, and a photo or drawing below.

Date ______________________ Time_________________________

Photo or Drawing

Observations	Wonderings

Name: ______________________________

Part 1

Changing Shadows

SAFETY NOTE: Do not look directly at the Sun for an extended period of time. Sunlight can cause eye damage.

1. When was your shadow the longest? ____________________

2. When was it the shortest? ____________________

3. How did the direction of your shadow change throughout the day?

____

____

4. How did the position of the Sun in the sky change throughout the day? ____________________

____

____

5. Why do you think your shadow changed throughout the day?

____

____

____

Part 2

Changing Shadows

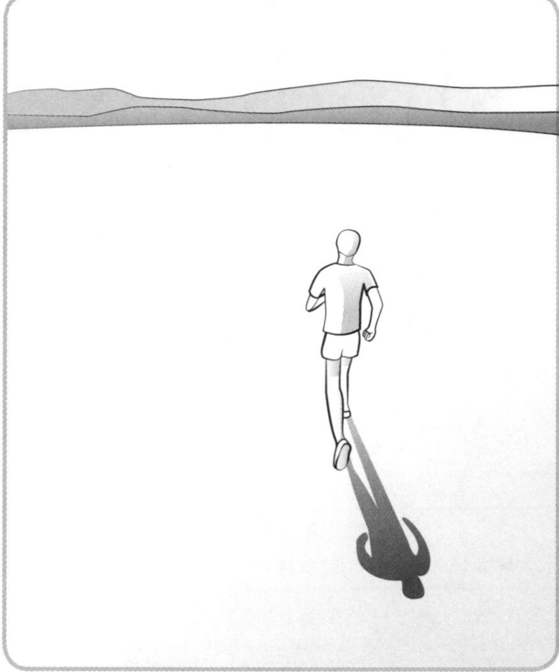

Compare the two pictures.

Do you think these two people are running at the same time of day?

How do you know? Explain your thinking.

Name: ____________________

Sunsets and Shadows Assessment

1. Why is it not scientifically accurate to say the Sun is "going down" at sunset?

____

____

____

____

____

____

____

____

____

____

____

____

2. If you go outside on a sunny day, you will see your shadow. Sometimes your shadow is longer than you are, and other times it is shorter than you are. How can this difference in the length of your shadow be explained? Use a labeled drawing to help explain your answer.

Problem Solvers

Description

Benjamin Franklin was a genius at solving problems. This lesson introduces Franklin's timeless inventions to inspire students to use the engineering design process to solve some of their own everyday problems.

Suggested Grade Levels: 3–5

LESSON OBJECTIVES *Connecting to the Framework*

ENGINEERING, TECHNOLOGY, AND APPLICATIONS OF SCIENCE

Core Idea ETS1: Engineering Design

ETS1.A: Defining and Delimiting an Engineering Problem

By the end of grade 5: Possible solutions to a problem are limited by available materials and resources (constraints). The success of a designed solution is determined by considering the desired features of a solution (criteria). Different proposals for solutions can be compared on the basis of how well each one meets the specified criteria for success or how well each takes the constraints into account.

ETS1.B: Developing Possible Solutions

By the end of grade 5: Research on a problem should be carried out—for example, through internet searches, market research, or field observations—before beginning to design a solution. An often productive way to generate ideas is for people to work together to brainstorm, test, and refine possible solutions. Testing a solution involves investigating how well it performs under a range of likely conditions. Tests are often designed to identify failure points or difficulties, which suggest the elements of the design that need to be improved. At whatever stage, communicating with peers about proposed solutions is an important part of the design process, and shared ideas can lead to improved designs.

There are many types of models, ranging from simple physical models to computer models. They can be used to investigate how a design might work, communicate the design to others, and compare different designs.

ETS1.C: Optimizing the Design Solution

By the end of grade 5: Different solutions need to be tested in order to determine which of them best solves the problem, given the criteria and the constraints.

Core Idea ETS2: Links Among Engineering, Technology, Science, and Society

ETS2.B: Influence of Engineering, Technology, and Science on Society and the Natural World

By the end of grade 5: Over time, people's needs and wants change, as do their demands for new and improved technologies. Engineers improve existing technologies or develop new ones to increase their benefits (e.g., better artificial limbs), to decrease known risks (e.g., seatbelts in cars), and to meet societal demands (e.g., cell phones). When new technologies become available, they can bring about changes in the way people live and interact with one another.

Featured Picture Books 

TITLE: ***Now & Ben: The Modern Inventions of Benjamin Franklin***
AUTHOR: **Gene Barretta**
ILLUSTRATOR: **Gene Barretta**
PUBLISHER: **Square Fish**
YEAR: **2008**
GENRE: **Narrative Information**
SUMMARY: *This colorfully illustrated picture book details the inventions and discoveries of Benjamin Franklin and how they are integrated into today's society.*

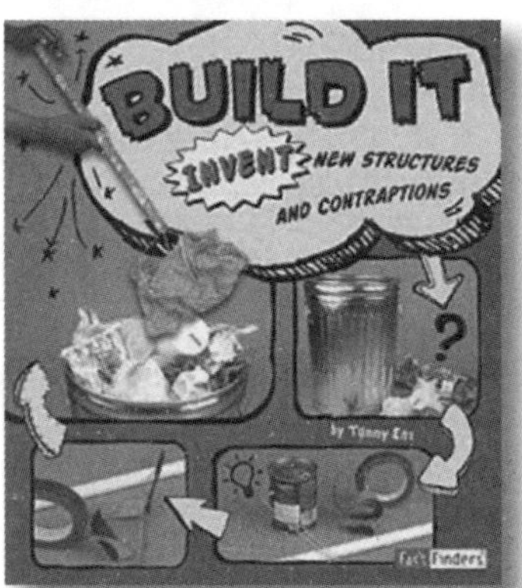

TITLE: ***Build It: Invent New Structures and Contraptions***
AUTHOR: **Tammy Enz**
PUBLISHER: **Capstone Press**
YEAR: **2012**
GENRE: **Non-Narrative Information**
SUMMARY: *Written by a civil engineer, this book takes students through the process of designing a solution to a problem. Eight projects are used as examples of the process.*

Time Needed

This lesson will take several class periods. Suggested scheduling is as follows:

Day 1: Engage with Faces on Our Money and *Now & Ben* Read-Aloud

Day 2: Explore with Ben Franklin's Inventions and Solving Problems

Day 3: Explain with *Build It* Read-Aloud

Day 4 and beyond: Elaborate with Designing Solutions and **Evaluate** with The Pitch

Materials

For Faces on Our Money (per class)

- Samples or pictures of U.S. currency (see *www.newmoney.gov/currency/images.htm* for downloadable images)

For Ben Franklin's Inventions

Per class

- Pictures of present-day versions of some of Ben Franklin's inventions, such as bifocals, lightning rod, long arm (extension arm), swim fins, Franklin stove, writing chair, clock or watch with a second hand, and odometer

Per team of three to four students

- 1 sheet of poster paper

For Solving Problems

- Problem Cards (for teacher to distribute)
- Poster paper
- Markers

For Designing Solutions

- Materials to be determined after lesson has begun

Student Pages

- Designing Solutions
- The Pitch
- The Pitch Scoring Rubric

Background

The process of inventing solutions to problems is the work of engineers. *A Framework for K–12 Science Education* suggests that students in grades K–6 be given opportunities to use engineering practices. These practices involve the application of scientific principles to solve problems. Even though science and engineering are two different fields, they go hand in hand. Look around the room—anything that is not part of nature was designed by someone. From the pen you write with, to the window you look through, to the cell phone in your pocket—all of these were designed by engineers. Engineering is sometimes referred to as the "stealth" profession because, although we use all of these designed objects, we seldom think about the engineering practices involved in their creation and production. It is important to not only give students an awareness of the work of engineers but to provide them with opportunities to think like engineers.

This lesson gets students thinking about solving problems by learning about a remarkable problem solver, Benjamin Franklin. He was a writer, printer, musician, cartoonist, humanitarian, statesman, entrepreneur, inventor, and—we would argue—engineer. Franklin had a creative mind and a strong desire to solve problems. Even though he lived more than 200 years ago, many of the inventions he designed are still being used today. After learning about Franklin and some of his timeless inventions, students will go through the engineering design process themselves by brainstorming everyday problems, researching and developing solutions to solve them, testing and revising the solutions, and, finally, pitching their ideas.

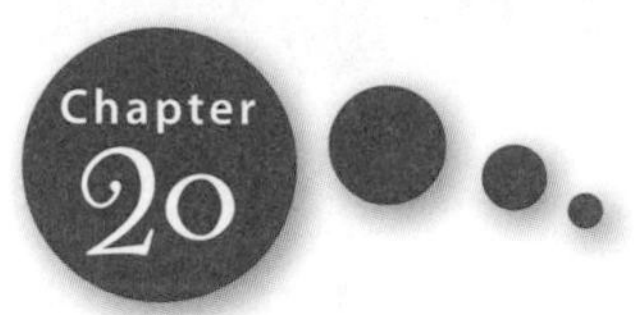

$100 BILL

engage

Faces on Our Money

Show students some samples or pictures of U.S. currency. *Ask*

- ? Do you know who the people are that are pictured on the money?
- ? Why do you think their pictures were printed on money?
- ? What would you have to do to deserve the honor of having your picture printed on money? (something good and important that affects a lot of people)
- ? Whose faces are on the $5(Abraham Lincoln), $20 (Andrew Jackson), and $50 (Ulysses S. Grant) bills?
- ? Why do you think they appear on our currency? (They were presidents.)
- ? Who is on the $100 bill? (Ben Franklin)
- ? Was Ben Franklin a president? (no)
- ? What did Ben Franklin accomplish in his lifetime that earned him a place on the $100 bill?

Now & Ben Read-Aloud

Determining Importance

Connecting to the Common Core
Reading: Informational Text
Key Ideas and Details: 3.1, 4.1, 5.1

Show students the cover of *Now & Ben.* Tell students that as you read the book aloud you would like them to think about why Ben Franklin deserved to have his portrait on the $100 bill.

After reading, *ask*

- ? Why do you think Ben Franklin earned a place on the $100 bill? (He made many discoveries, created many inventions, and created community systems such as libraries, post offices, fire departments, and hospitals. He also had an important role in writing many important documents, including the Constitution and the Declaration of Independence. As the book states, "It is remarkable that one man could achieve so much in a lifetime" (p. 31).
- ? Which of Franklin's accomplishments impressed you the most? Why? Have students turn and talk with a partner.

explore

Ben Franklin's Inventions

Tell students that you would like to focus on some of the inventions of Ben Franklin. *Ask*

- ? What is an invention? (something that is made to meet a need or solve a problem)
- ? What is the difference between an invention and a discovery? (An invention is something that is created; a discovery is something that is found for the first time.)
- ? What discoveries did Ben Franklin make? (He discovered that lightning was electric current,

BIFOCALS WERE INVENTED BY BEN FRANKLIN

that citrus fruits help prevent a disease called scurvy, and that the Gulf Stream helped ships travel faster across the Atlantic Ocean.)

? What are some of the inventions of Ben Franklin? (bifocals, lightning rod, long arm, fins for swimming, glass armonica, Franklin stove, writing chair, rocking chair with a fan, rocking chair that churned butter, odometer)

Show students photos or actual examples of the inventions of Benjamin Franklin. For each invention, discuss the problem that each invention solved.

Ask

? Have you ever invented something?

Solving Problems

Tell students that inventions are designed to solve problems, and you have some problems you would like them to solve in teams. Give each team of three to four students one of the problem cards. Have them read the problem, then explain that research on a problem should be carried out—for example, through internet searches, market research, or field observations—before beginning to design a solution. Have them do a brief internet search on some of the ways this problem or related problems have been solved by others. Then have each team brainstorm ideas for solving it. They should choose their best idea and design it on a sheet of poster paper. Have them come up with a name for the invention, label its parts, and list the materials needed to make it. Have students share their designs in a poster session.

explain

Build It Read-Aloud

Tell students that you have a nonfiction book about inventing that offers solutions to the eight problems they worked on solving. Show them the cover of *Build It*. Read the "About the Author" section on the last page of the book so the students can learn more about Tammy Enz, a civil engineer who loves to tinker and figure out how things work. She also loves to read.

Making Connections: Text-to-World

Ask

? What do engineers do? (design solutions to problems)

? Do you know anyone who is an engineer? What kind?

? What kinds of problems do they solve?

Read page 4, and explain that, all together, the six steps of inventing can also be referred to as the *engineering design process*. Point out that there are many variations of the engineering design process, but they all achieve the same goal of solving a problem.

Questioning

> Connecting to the Common Core
> **Reading: Informational Text**
> Integration of Knowledge and Ideas: 3.7, 4.7, 5.7

As you read each of the six steps on page 5, ask the following questions about the teams' designs:

Step 1: Problem

? What was the problem you were trying to solve? Turn and talk.

Step 2: Principle

? What scientific rules or laws apply to the problem your team was trying to solve? Turn and talk.

Step 3: Ideas

? What did you do before your team brainstormed ideas? (We did internet research on other solutions that have been designed to solve the problem or similar problems.)

? What were some of the ideas your team brainstormed? Why did you choose the one you did?

? How did your research influence the solution you chose? Turn and talk.

Step 4: Plan

? What tools and supplies would you need to actually build the device? Turn and talk.

Step 5: Create

? Do you think you could actually build the invention you designed? Turn and talk.

Step 6: Improve

? How would you test your invention? How do you think it could be improved? Turn and talk.

After reading page 5, *ask*

? Does the process end after the invention is created? (No, it ends with improving the invention or starts over with a new problem.)

The rest of the book describes solutions to the problems the students worked to solve. Read the first page of each project aloud, then share the picture of the completed inventions. As they listen, have them compare their solutions with the ones in the book. Be sure to read the additional information on pages 8, 13, 19, and 27, which tell the stories of how various inventions came to be and explain the significance of a patent.

elaborate

Designing Solutions

Depending on time and materials, this activity can be done in two different ways:

1. The students and teacher can all work on a solution to the same problem. This will require a smaller amount and less diverse range of resources and materials. This would be ideal if you are planning to do the entire activity in the classroom without any home time. If you choose this option, we suggest you brainstorm the problems together and decide on a problem to which all students can relate.
2. Each student or pair of students chooses their own problem to solve. This would be a good option if the majority of the work is to be done at home, but it will require a larger amount and more diverse range of resources and materials that parents may have to provide.

Tell students that they are going to have the opportunity to apply what they have learned about designing solutions to problems by inventing a solution to a problem that they have. Begin by having students brainstorm problems that they have at school and home and listing them on the Designing Solutions student page. Explain that the parts of the inventing process that were featured in the book *Build It* are listed as steps on these pages. Have students use the student pages to help them follow the design process, stopping at the teacher checkpoints to discuss their ideas with you and get your signature at each checkpoint before moving on. It is up to you which parts of this process are done at school and which parts are done at home. Here are some suggestions for how to support students in each phase of the design process:

- *Problem:* Encourage students to think about the tasks they encounter on a regular basis at school or at home.
- *Principle:* Students may need help on this section to bring in the scientific principles that are part of designing their product/invention. You can help by providing books, a trip to the library, and/or internet access to help them research their problem.
- *Ideas:* When brainstorming ideas, it is helpful to find out what invention/product already exists to solve this problem. Internet research is a good way to find out.
- *Plan:* Read through students' plans carefully before signing off on them. Make sure their plan is realistic and safe. This is the point where they really need guidance if they are going to actually create their product/invention.
- *Create:* If you plan to have them create their products/inventions in class, collect the appro-

priate materials and have them ready.

- *Improve:* Explain that an important part of inventing is testing your invention/product so you can evaluate how it works. Tell students that inventions/products can always be improved. If their invention/product actually works, students may struggle to find a way to improve it. If this is the case, ask them to think about how it could be made stronger, cheaper, more fun, or even better looking.

The Pitch

> Connecting to the Common Core
> **Speaking and Listening**
> Presentation of Knowledge and Ideas: 3.4, 4.4, 5.4

Tell students that after solutions are designed and inventions are created and improved, the next step is usually pitching the invention to a company or investor that can help manufacture and sell it. Tell students that you would like them to "pitch" their inventions to you. This can be presented live or through a recorded commercial. Students can also create visual aids, such as posters, props, or slide shows to use in their pitch. You may want to show students some clips from the ABC television show *Shark Tank* (available at *http://abc.go.com/shows/shark-tank)* where entrepreneurs who have designed solutions to problems must make their pitches. This will give students some inspiration and direction when preparing to pitch their product/invention. Pass out both The Pitch student page and The Pitch Scoring Rubric to help students plan their pitch, and use the rubric to evaluate students.

Websites

Discover Engineering
www.discoverengineering.org

Franklin Institute: Benjamin Franklin
http://sln.fi.edu/franklin/inventor/inventor.html

PBS: Benjamin Franklin
www.pbs.org/benfranklin

PBS Kids: Zoom Into Engineering
www.pbs.org/parents/zoom/engineering

More Books to Read

Casey, S. 2005. *Kids inventing! A handbook for young inventors.* Hoboken, NJ: Jossey-Bass.
Summary: This book is for kids who are serious about turning their ideas into profit. Explains all the steps of inventing, including filing for a patent, selling your idea, and so on.

Enz, T. 2012. *Harness it: Invent new ways to harness energy and nature.* North Mankato, MN: Capstone Press.
Summary: This book describes various ways to harness energy, including several different projects that the reader can create.

Enz, T. 2012. *Repurpose it: Invent new uses for old stuff.* North Mankato, MN: Capstone Press.
Summary: This book describes how to invent new things and save the environment at the same time. Includes several projects the reader can create.

Enz, T. 2012. *Zoom it: Invent new machines that move.* North Mankato, MN: Capstone Press.
Summary: This book describes the process of inventing and includes several sample moving inventions that the reader can create.

Satterfield, K. H. 2005. *Benjamin Franklin: A man of many talents.* New York: HarperCollins.
Summary: This *Time for Kids* biography details the life of Benjamin Franklin.

Inquiry Place

Have students brainstorm questions about inventions. Examples of such questions include

? Can you build one of the inventions from the *Build It* book? Try it!

? What do kids your age think is the best invention of all time? Survey it!

? How did a common invention you use every day come to be? Research it!

? How do you apply for a patent? Research it!

Then have students select a question to research as a class, or have groups of students vote on the question they want to research as a team. Have students present this information on a poster. Students will then share their findings in a poster session or gallery walk.

Problem Cards

(from *Build It: Invent New Structures and Contraptions* by Tammy Enz)

Getting out of bed to open or close your bedroom door is a real pain. How can you build something that will open and close the door from across the room?

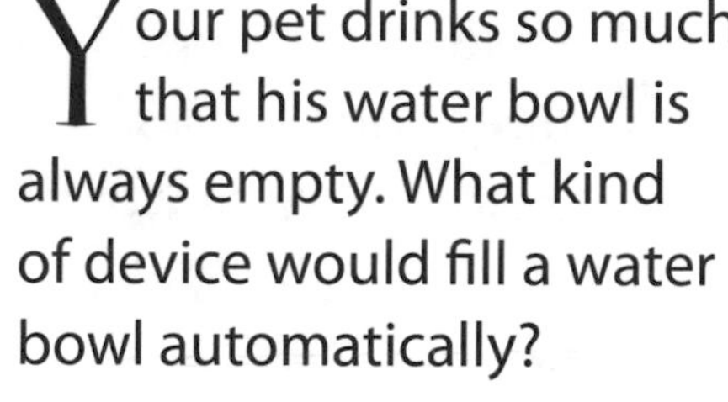

Your pet drinks so much that his water bowl is always empty. What kind of device would fill a water bowl automatically?

Newspapers pile up day after day. There must be something fun to do with them. Could newspapers be used to build a fort?

It's a bother getting out of the pool to get something to drink. How could you stay afloat and quench your thirst at the same time?

Problem Cards **Continued**

Garbage clutters parks, streets, and lawns. What kind of device would be most helpful to pick up the trash?

Eggs crack with the slightest bump. What kind of contraption will protect an egg that is dropped from a large height?

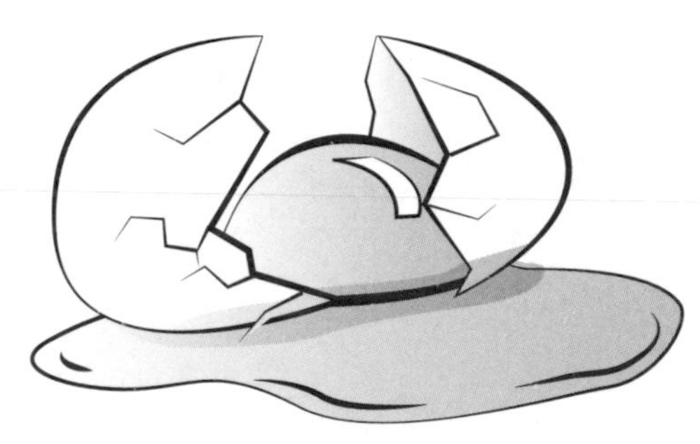

Spare change has a way of collecting in wallets, purses, and odd containers. What kind of machine could help sort your mixed coins?

Toothpicks are great for getting broccoli out of your teeth. But do they have other uses? Could they be used to build a bridge?

Name ____________________

Designing Solutions

List some problems you have in your everyday life. Think about which problems could be solved with an invention. Then circle the one that you will try to solve.

What scientific rules or laws apply to the problem you are trying to solve?

Teacher Check ____________________

What ideas could help you solve the problem? Research some of the ways this problem or related problems have been solved by others. Then list your own ideas.

What supplies will you need and how will you build your solution?

STOP

Teacher Check ______________________

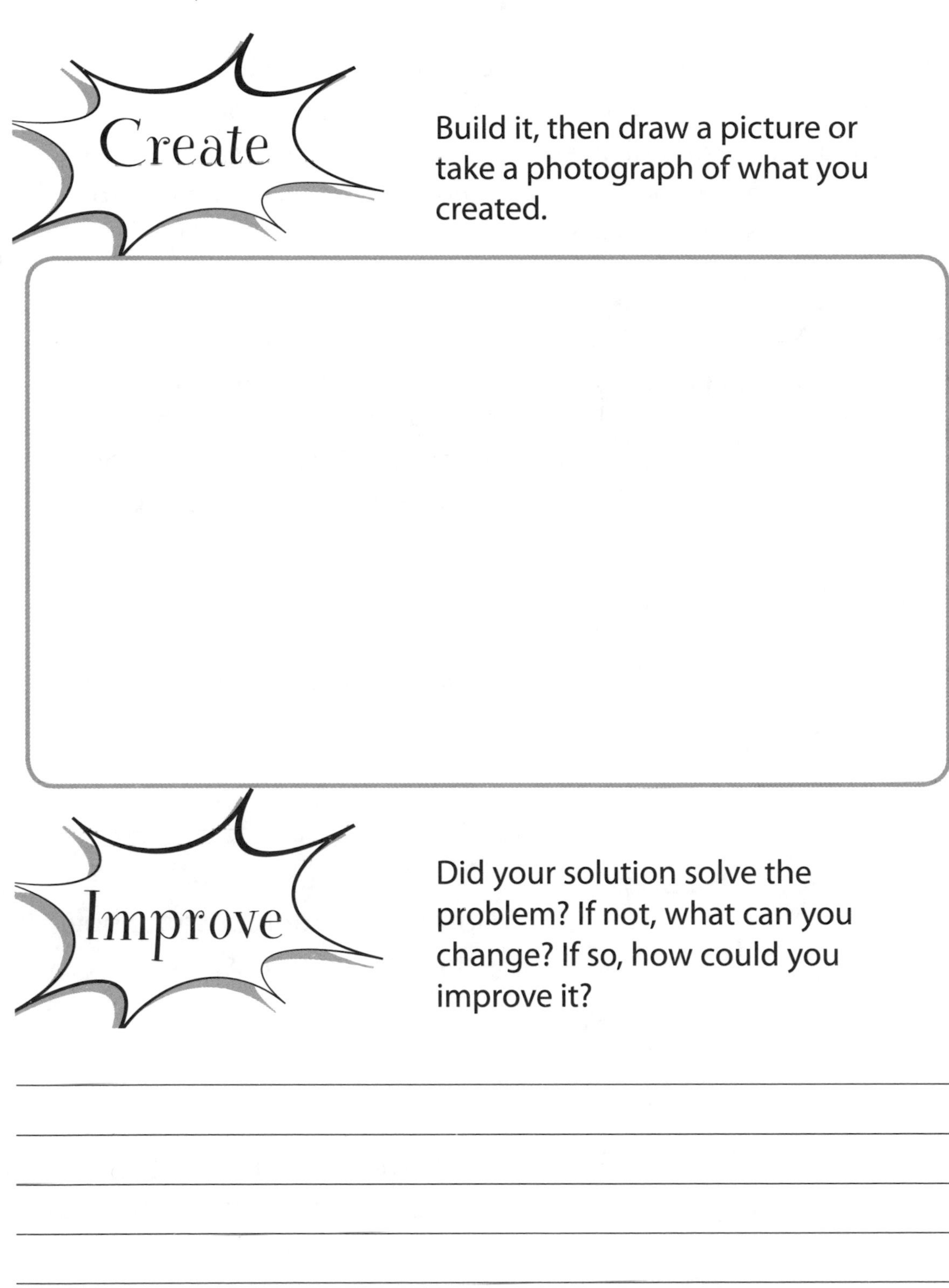

Build it, then draw a picture or take a photograph of what you created.

Did your solution solve the problem? If not, what can you change? If so, how could you improve it?

Name ______________________________

The Pitch

Imagine that you are trying to pitch your invention to a company or investor that can help you manufacture and sell it. Your pitch can be presented live or through a recorded commercial. You can use visual aids, such as posters, props, or slide shows, to use in your pitch. You must include these four elements:

What do you call your invention? Does the name clearly communicate what your invention does? How will you display the name in your pitch?

What problem does your invention solve? Who has this problem?

How will you demonstrate how your product works?

Are there similar products on the market? How will you prove that this product is better?

Name : ______________________________

The Pitch Scoring Rubric

Imagine that you are trying to pitch your invention to a company or investor that can help you manufacture and sell it. Your pitch can be presented live or through a recorded commercial. You can use visual aids in your pitch, such as posters, props, or slide shows, to use in your pitch. You must include these four elements:

4-Excellent **3-Above Average** **2-Average** **1-Below Average**

Score	Criteria
___ 4 ___ 3 ___ 2 ___ 1	***Name*** Name of invention communicates what the invention does. Name is clearly displayed or announced.
___ 4 ___ 3 ___ 2 ___ 1	***Problem*** The problem that is solved by the invention is clearly described. The people who are potentially affected by the problem are identified.
___ 4 ___ 3 ___ 2 ___ 1	***Demo*** The invention is demonstrated in a clear and effective manner.
___ 4 ___ 3 ___ 2 ___ 1	***Comparison*** Similar products on the market are identified. Product is shown to be better than similar products on the market.

___ Total Points/16

Index

Page numbers printed in **boldface** type indicate tables or illustrations.